Biochemical Responses Induced by Herbicides

Biochemical Responses Induced by Herbicides

Donald E. Moreland, EDITOR
U.S. Department of Agriculture

Judith B. St. John, EDITOR
U.S. Department of Agriculture

F. Dana Hess, EDITOR
Purdue University

Based on a symposium
sponsored by the Division
of Pesticide Chemistry
at the 181st ACS National
Meeting, Atlanta, Georgia,
March 29–April 3, 1981.

ACS SYMPOSIUM SERIES 181

AMERICAN CHEMICAL SOCIETY
WASHINGTON, D. C. 1982

Library of Congress CIP Data

Biochemical responses induced by herbicides.
(ACS symposium series, ISSN 0097-6156; 181)

Includes bibliographies and index.

1. Plants, Effect of herbicides on—Congresses. 2. Herbicides—Physiological effect—Congresses.
I. Moreland, Donald E., 1919- . II. St. John, Judith, 1940- . III. Hess, F. Dana, 1946- . IV. American Chemical Society. Division of Pesticide Chemistry. V. Series.

QK753.H45B56 581.2'4 81-20645
ISBN 0-8412-0699-6 AACR2 ACSMC8 181 1-274
1982

PRINTED IN THE UNITED STATES OF AMERICA

ACS Symposium Series

M. Joan Comstock, *Series Editor*

FOREWORD

The ACS SYMPOSIUM SERIES was founded in 1974 to provide a medium for publishing symposia quickly in book form. The format of the Series parallels that of the continuing ADVANCES IN CHEMISTRY SERIES except that in order to save time the papers are not typeset but are reproduced as they are submitted by the authors in camera-ready form. Papers are reviewed under the supervision of the Editors with the assistance of the Series Advisory Board and are selected to maintain the integrity of the symposia; however, verbatim reproductions of previously published papers are not accepted. Both reviews and reports of research are acceptable since symposia may embrace both types of presentation.

CONTENTS

PREFACE

The symposium at which the papers in this volume were presented provided an opportunity to present, discuss, and publish a comprehensive account of the current status of research concerned with basic biochemical and physiological responses associated with the phytotoxic action of herbicides.

Several of the contributions consider interferences by herbicides with photoinduced electron transport in chloroplasts. These papers include such topics as sites of interference on the electron transport pathway, identification and biogenesis of the photosystem II (PS II) triazine-binding protein and lipid composition in triazine-susceptible and triazine-resistant interactions between herbicides and the PS II binding site, the relation between interference with photoinduced electron transport and the expression of phytotoxicity, differences between properties of the triazine-binding protein and lipid composition in triazine-susceptible and triazine-resistant biotypes of weed species, and alterations to the permeability properties and lipid composition of organelle membranes induced by herbicides, some of which are PS II inhibitors. Other contributions report on the activation of diphenyl ether herbicides by light, effects of herbicides and growth regulators on pigment metabolism, the complex metabolic alterations induced by glyphosate, the use of unicellular green algae to differentiate between physiological modes of herbicidal action, and procedural details that can be used to differentiate between effects imposed on cellular division and cellular enlargement.

As evidenced by the contents of the papers, considerable progress has been made in recent years in the identification of mechanisms and modes of action of herbicides. However, the elucidation of the mechanisms of action of herbicides at the cellular and molecular levels, together with the translation of the primary perturbations to the mode of action that leads to the expression of phytotoxicity, will continue to challenge the ingenuity of investigators for many years. A clear insight into the physiological and biochemical mechanisms through which herbicides operate depends on the development of a better comprehension of growth, and the factors through which it is controlled, at the molecular level. Each year background information increases, as does knowledge of the regulation of subcellular metabolism, which helps clarify the action of herbicides. Consequently, future symposia can be expected to witness the report of substantial and exciting progress.

We would like to acknowledge the cooperation and enthusiasm provided by the speakers with their presentations, their participation in the symposium, and in the development of their manuscripts. We are especially grateful for the assistance in editing, typing, and finalization of the manuscripts that was provided by Linda M. Lynch and William P. Novitzky.

DONALD E. MORELAND
U.S. Department of Agriculture
Agricultural Research Service
Departments of Crop Science and Botany
North Carolina State University
Raleigh, NC 27650

JUDITH B. ST. JOHN
U.S. Department of Agriculture
Beltsville Agricultural Research Center
Beltsville, MD 20705

F. DANA HESS
Purdue University
Department of Botany and Plant Pathology
West Lafayette, IN 47907

October 13, 1981

1

Photosystem II Inhibiting Chemicals

Molecular Interaction Between Inhibitors and a Common Target

C. J. VAN ASSCHE and P. M. CARLES

ROUSSEL UCLAF, C.R.B.A. PROCIDA, Crop Protection, Scientific Division, St. Marcel, F. 13011 Marseille, France

From several lines of evidence (fluorescence and chemical-triggered luminescence measurements, competition studies, and mild trypsin digestion) various chemical families of PS II inhibitors appear to act the same way and upon a common site associated with the reducing side of photosystem II and the B protein, a specialized plastoquinone-protein complex. A model, using molecular orbital approaches, is proposed that is consistent with a hypothesis for the existence of hydrogen bonds between inhibitors and a simulated proteinaceous target. Short- and long-range intermolecular energies within the micro-environment of the "receptor" B - protein complex show that in calculating reciprocal interactions between inhibitors and the simulated binding site, polarization and dispersion components are relatively more important than electrostatic energy.

Many commercially available herbicides have been demonstrated to interfere with one or more steps of photosynthesis, by reacting near the photosystem II (PS II) center [for a recent review see (1), among others]. DCMU (2) and other chemical families of photosynthetic inhibitors (3, 4) were shown to shift the potential of the PS II secondary electron acceptor B, a specialized plastoquinone molecule, bound to a protein (5).

PS II inhibitors bind to a hydrophobic domain of a 32 Kdalton polypeptide considered either to be the B-protein or to be associated closely with it (6). The exact nature of the molecular interaction between inhibitors and target has not been elucidated; however, a model was proposed, using a molecular orbital approach (7), in which support was given to the postulate that hydrogen bonds were involved in the interaction between inhibitors and a simulated target, i.e., a dipeptide unit. This paper will describe results obtained with the approach presented

0097-6156/82/0181-0001$05.25/0

previously (7) for DCMU-type inhibitors and for compounds that exhibit a different site of action. These studies were conducted to determine if the compounds interfered with the same molecular target. A multimethodological approach will be described herein for this purpose.

Evidence of a Common Molecular Target for Various Inhibitor Chemical Families

Indirect Measurements of Effects on PS II Primary Electron Acceptor. When photosynthetic organelles (microalgae or chloroplasts) are illuminated, electrons extracted from water are transported to a final electron acceptor $NADP^+$ which becomes reduced. Intermediary electron acceptors of the photosynthetic electron transport chain undergo various redox transients. This chain may be represented simply as follows:

$$H_2O \rightarrow S_0 \rightarrow S_1 \rightarrow S_2 \rightarrow S_3 \rightarrow S_4 \rightarrow PS\ II \rightarrow Q \rightarrow B \rightarrow PQ \rightarrow PS\ I \rightarrow NADP^+$$

$$S_4 \rightarrow S_0 + O_2$$

Chlorophyll a_{II} fluorescence yield will depend not only on the redox state of the PS II primary electron acceptor Q, but also on the donor side Z of PS II, from which 4 oxidizing equivalents are required for the evolution of O_2 from water, and the PS II secondary electron acceptor B. Therefore, fluorescence represents a meaningful probe for testing interference with the redox state of the PS II complex.

Chemicals listed in Table I have been tested for their effects on chlorophyll fluorescence. All chemicals show prompt fluorescence kinetics similar to that of DCMU, except for nitrofen and trifluralin which are excluded from the group of so-called "DCMU-type inhibitors". A quantitative relationship between inhibitor concentration and selected fluorescence parameters has been researched.

Steady-state fluorescence or Φ_{max} (see Figure 1) is reached when Q is completely reduced, at least in *Chlorella*. Half-effect concentrations for each DCMU-type inhibitor, expressed as the concentration that gives rise to a 50% increase of Φ_{max} and termed $p\Phi_{max50}$, are shown in Table II. The pI_{50} and $p\Phi_{max50}$ values are of the same order of magnitude, although Φ_{max} values are, on the average, higher than values for inhibition of the Hill reaction. These differences might be caused by experimental conditions that are not identical in both cases (chlorophyll concentration, electron acceptor, light intensity, etc.). Therefore, other fluorescence parameters have been studied in order to find a better correlation.

Murata et al., (8) proposed that the area A_{max} (Figure 1) over the fluorescence rise normalized to Φ_{max}, was proportional

TABLE I

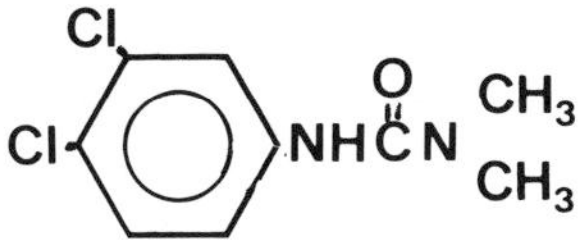

DCMU *(N-phenylurea)*	Barban *(N-phenylcarbamate)*
Atrazine *(s-triazine)*	Propanil *(N-acylanilide)*
Metribuzin *(triazinone)*	Lenacil *(uracil)*
Ioxynil *(hydroxybenzonitrile)*	Pyramin *(pyridazinone)*
Trifluralin *(dinitroaniline)*	RU 21731 *(carbamoyl-thiadiazoline)*
Nitrofen *(diphenyether)*	I-dinoseb *(dinitrophenol)*

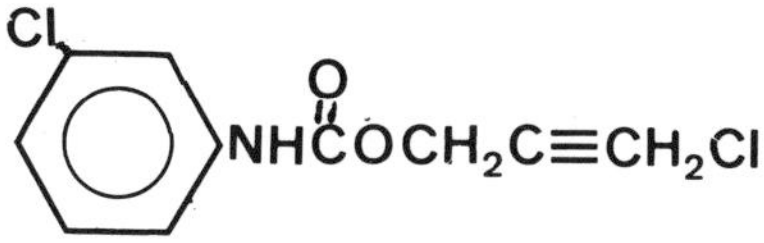

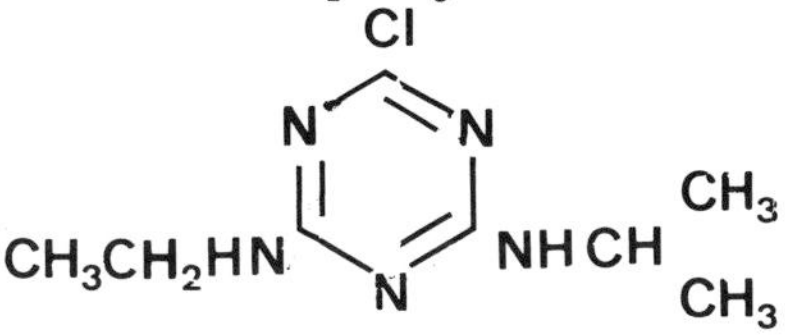

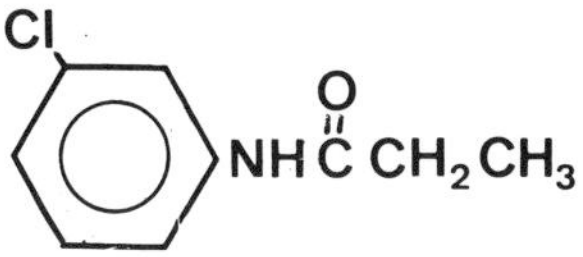

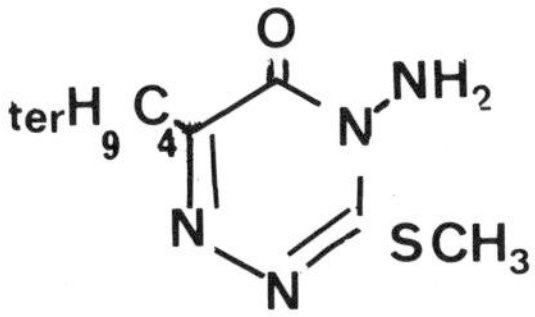

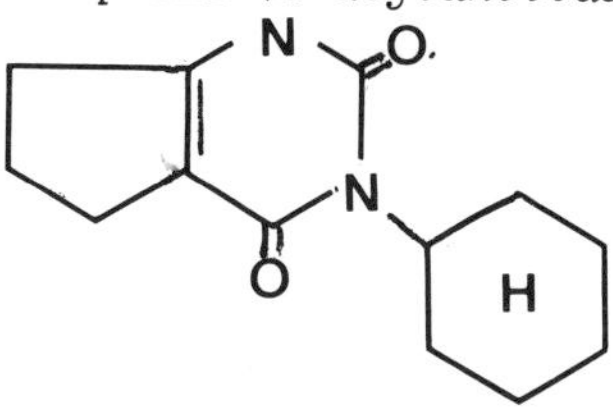

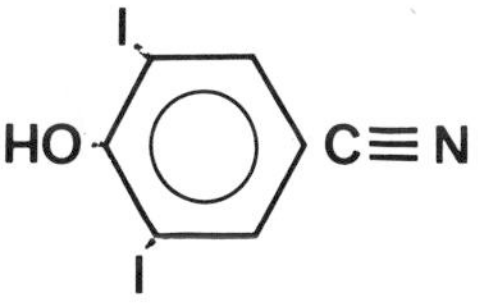

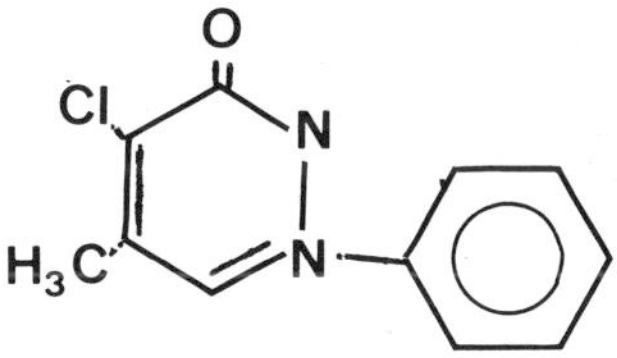

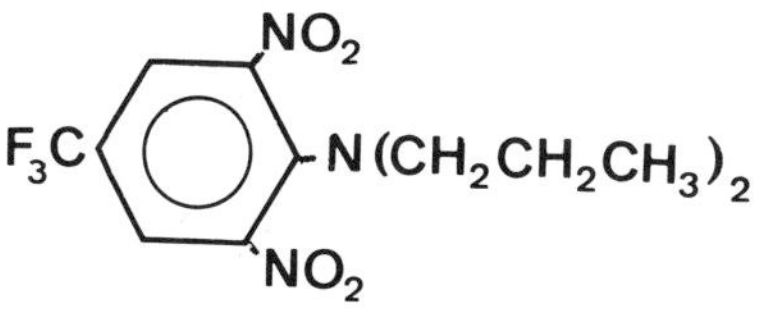

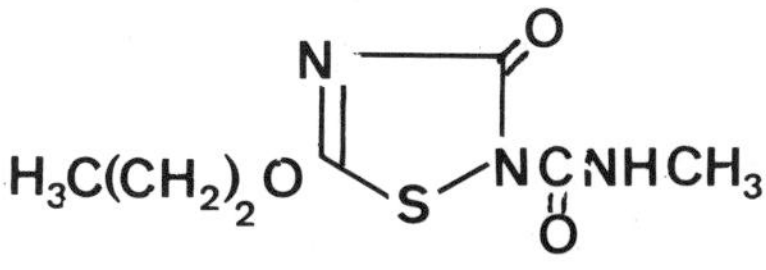

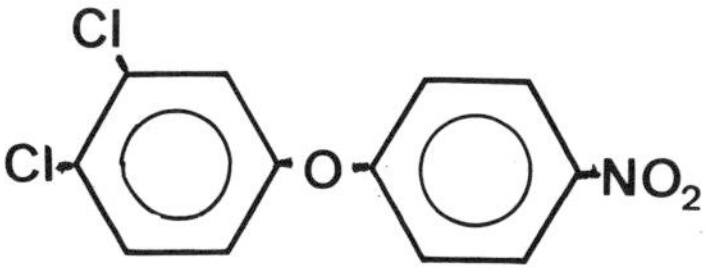

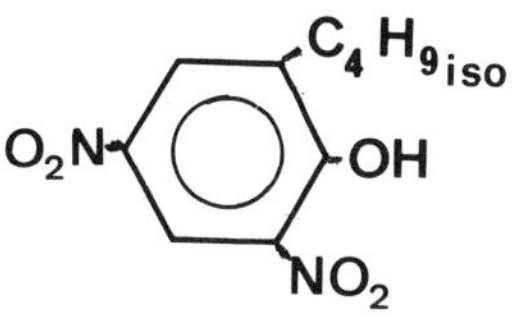

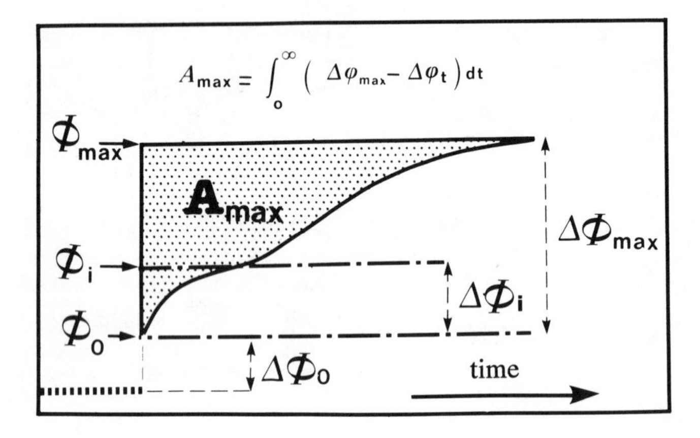

Figure 1. Definitions of the different fluorescence parameters on chloroplasts of Chlorella pyrenoidosa. *Fluorescence intensity is expressed in arbitrary units.*

TABLE II
Comparison between various chlorophyll a_{II} fluorescence parameters and inhibition of photosynthetic electron transport measured on *Chlorella* and chloroplasts with PS II inhibitors

Compound	pI_{50} Hill reaction		$p\Phi_{max\ 50}$	$pA_{max\ 50}$		$p\Phi_{v50}$	$p\Phi o_{50}$
	Chlorella	Chloroplast	Chlorella	Chlorella	Chloroplast	Chloroplast	Chloroplast
DCMU	6.82	6.78	7.22	6.8	6.7	6.62	6.10
Barban	4.00	4.05	5.62	4.00	5.2	4.48	3.31
Propanil	5.29	6.15	6.22	5.31	6.07	6.15	5.04
Atrazine	5.75	6.49	6.92	5.68	6.34	6.38	5.12
Lenacil	6.67	6.23	6.40	6.58	6.69	6.46	6.10
Pyramin	5.85	5.43	5.22	5.82	5.52	5.30	5.09
Ioxynil	6.57	6.20	7.30	6.01	5.70	5.65	4.45
Metribuzin	6.57	6.40	7.09	6.49	6.35	6.18	6.04
RU 21731	5.55	6.30	6.53	5.45	6.20	6.26	0

to the number of quanta used by the PS II reaction center. In other words, A_{max} represents the size of the PS II electron acceptor pool (9) or the photochemical quenching capacity of this pool (10). A_{max} varies with inhibitor concentration so that a half-effect value can be calculated and termed as $pA_{max_{50}}$ (cologarithm of the concentration giving 50% A_{max} stimulation in *Chlorella* or decrease in chloroplasts). These values are shown in Table II and correlate very well with pI_{50}'s for all inhibitors except ioxynil.

During fluorescence induction, a first transient, Φi, is observed that corresponds to the oxidation of an intermediate pool and Q by PS I (11). This fluorescence transient can be measured easily for chloroplast suspensions in the presence of inhibitors (Figure 1). From the values observed on the induction curves, the corresponding $\Delta\Phi_{max} - \Delta\Phi_{i} / \Delta\Phi_{max}$ were calculated for each inhibitor concentration from which the cologarithm of concentrations of half-effect were calculated, termed $p\Phi_{v50}$, and shown in Table II. Again, a good correlation exists between fluorescence transient data and inhibition of the Hill reaction, except for ioxynil.

Direct Interferences on the PS II Secondary Electron Acceptor, B. DCMU shifts the potential of the PS II secondary electron acceptor B to a more electronegative value, such that Q is reduced by B, resulting in a block of the reoxidation of Q. Several techniques were developed to test the ability of inhibitors described herein to displace the following equilibrium to the right:

$$QB^{-} \rightleftharpoons Q^{-}B$$

The initial level of fluorescence Φ_0 (Figure 1) corresponds to a physiological state in which all of the interconnecting systems between PS II and PS I are oxidized. Among the factors that affect Φ_0, both the system of oxidizing charge accumulation and the PS II secondary electron acceptor B were demonstrated to play an important role (12).

Because the redox state of the complex ZQB controls fluorescence yield, Z must be inactivated in order to probe QB interactions. A pretreatment of chloroplasts with high concentrations of hydroxylamine will completely destroy the oxidizing side of PS II (Z) and produce some reduction of B (13). The experimental protocol was the following: first, a chloroplast suspension was treated in the dark with 10 mM hydroxylamine and Φ_0 was measured immediately. After a 5 min dark incubation period, the chloroplasts received a strong actinic illumination that gave rise to the QB^- form. After complete relaxation (15 min), DCMU-type inhibitors were added producing Q^-B and Φ_0 was measured again. The difference between the two values was evaluated in relation to the inhibitor concentrations and a half-effect determined.

This value, termed p Φ_{050}, is compared to other fluorescence data and to pI_{50}'s on the Hill reaction (Table II).

Although p Φ_{050} values are slightly smaller than pI_{50}'s obtained for the Hill reaction, they remain very similar except for ioxynil. Generally speaking, quantitative measurements of the fluorescence parameters described herein showed that DCMU-type inhibitors disconnect the two photosystems progressively as concentrations of the inhibitors are increased. Ioxynil does not behave in this manner. Ioxynil may not interfere at exactly the same site as the DCMU-type inhibitors, or it may have a secondary point of interference with the photosynthetic electron transport chain.

Chemical Triggered-Luminescence. The quantum conversion within the photosynthetic apparatus results from the photochemical functioning of a "vectorial" system that uses light energy collected by an antenna chlorophyll molecule to produce a charge separation of an electron and a hole (14). Luminescence arises from a charge recombination between hole and electron. In dark-adapted material, luminescence can be triggered by several means, including certain chemicals. DCMU (15) and DCMU-type inhibitors (16) trigger luminescence of dark-adapted *Chlorella* in which some positive (Z^+) and negative (B^-) charges are stored, according to the following scheme:

$$Z^+ \ \mathrm{Chl} \ QB^- \xrightarrow{\text{DCMU-type compounds}} Z^+ \ \mathrm{Chl} \ Q^-B \rightarrow Z \ \mathrm{Chl} \ QB + h\nu$$

The resulting "dark" luminescence, can be measured with the simple experimental set-up previously described (15). A typical recording of chemical triggered luminescence is shown in Figure 2.

Most of the DCMU-type inhibitors trigger luminescence in the same way, i.e., there is a fast emission of light (L_i) followed by a slow relaxation (several min) back to the level of background luminescence (L_0) (Figure 2). However, there are two exceptions. RU 21731 a thiadiazoline, for which relaxation follows complex kinetics that cannot be explained readily. Ioxynil and *i*-dinoseb do not trigger luminescence; in fact, background luminescence is eliminated progressively (Figure 2).

In addition, H_2O_2-triggered luminescence is inhibited by *i*-dinoseb, ioxynil, and hydroxylamine, indicating a destruction or a deactivation of the donor side of PS II (17). This result would also explain the apparent discrepancy between fluorescence and luminescence data. Consequently, both ioxynil and *i*-dinoseb could affect a second site of action on the oxidizing side of PS II, because these compounds showed, in fluorescence experiments a qualitative response similar to that of DCMU-type inhibitors (not shown here for *i*-dinoseb).

When silicomolybdic acid is used as electron acceptor, O_2 evolution from illuminated chloroplasts becomes almost insensitive to DCMU (18) and DCMU-like inhibitors (4, 16). Ioxynil and

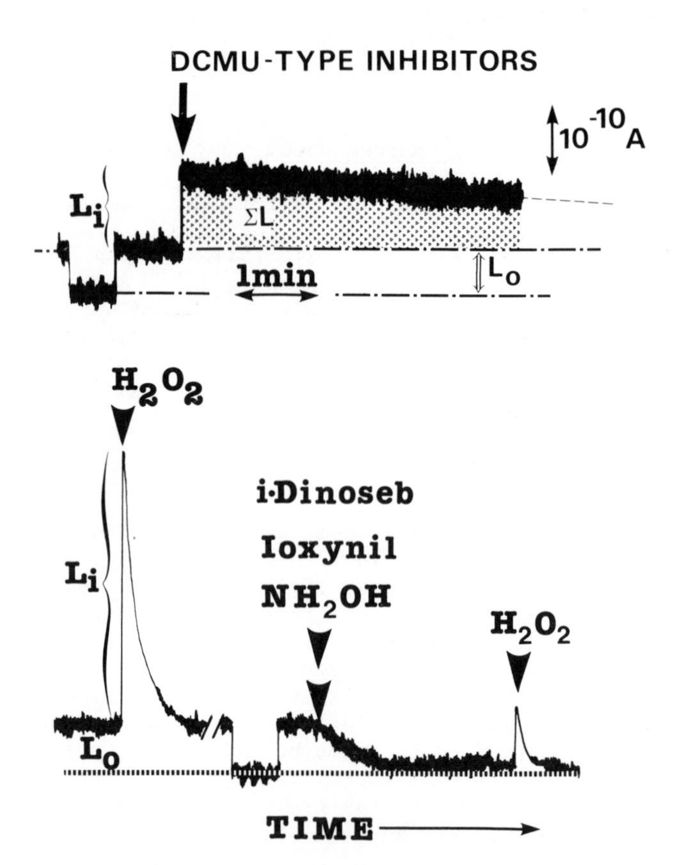

Figure 2. Trace recordings of photosynthetic inhibitor-triggered luminescence in Chlorella.

i-dinoseb were tested on DCMU insensitive silicomolybdate O_2 evolution: results show clearly that these two compounds interfere with electron transport near the oxygen evolving system.

Trifluralin and nitrofen also do not trigger luminescence: this observation confirms the assumption of a site of action different from DCMU for these two compounds.

A quantitative relationship between concentration of DCMU-type inhibitors and luminescence data can be established: cologarithm of molar concentration producing 50% increase of luminescence initial peak heights (pLi_{50}) and time integrals ($p \sum L_{50}$), respectively, are compared with pI_{50}'s in Table III. A good correlation between luminescence parameters and inhibition of electron transport exists, which indicates that the DCMU-type inhibitors shift the redox potential value of B to a more electronegative value. This may be responsible for inhibition of electron transport.

TABLE III
Comparison between half-effect values of luminescence data and pI_{50}'s for various photosynthetic inhibitors.

Compound	Luminescence initial peak heights (pLi_{50})	Luminescence integrals ($p \sum L_{50}$)	Electron transport (Ferricyanide) (pI_{50})
DCMU	6.80	6.80	6.82
Propanil	6.2	6.2	5.62
Pyramin	6.35	5.6	5.85
Atrazine	5.9	5.8	5.76
Lenacil	6.45	6.3	6.67
Metribuzin	6.8	6.9	6.57
Barban	4.8	4.84	4.0
Ioxynil	0	0	6.57
RU 21731	5.40	?	5.55
Nitrofen	0	0	4.25
Trifluralin	0	0	4.85

Interferences of Photosynthesis Inhibitors with a Common Chloroplastic Membrane Protein

The PS II electron acceptor pool has been shown to be covered by (19) or associated with (20) a proteinaceous shield that is necessary for electron transport as well as for sensitivity to inhibitors.

Mild-Trypsin Digestion. When a preparation of stripped chloroplasts (thylakoids) is incubated with trypsin, the electron transport block imposed by photosynthetic inhibitors is almost completely overcome. Based on responses obtained, inhibitors can be divided into 3 groups (Table IV). Chemicals in Group 1 behave like DCMU, i.e., most of the reducing activity is restored in trypsin-treated chloroplasts relative to untreated controls (*N*-phenylureas, *s*-triazines, *N*-acylanilides, *N*-phenylcarbamates, uracils, pyridazinones, and triazinones). Ioxynil and *i*-dinoseb

TABLE IV
Effect of various photosynthesis inhibitors on reducing activity of trypsin-treated chloroplasts.

	Treatment	Fe^{+++} reducing activity (% control)	Fe^{+++} reducing activity (% trypsin control)
Group 1	Control	100	--
	DCMU 10^{-6} M	25	--
	Trypsin	50	100
	Trypsin + DCMU 10^{-6} M	45	90
Group 2	Control	100	--
	Ioxynil 10^{-6} M	25	--
	Trypsin	55	100
	Trypsin + Ioxynil 10^{-6} M	27	49
Group 3	Control	100	--
	Nitrofen 10^{-4} M	40	--
	Trypsin	56	100
	Trypsin + Nitrofen 10^{-4} M	21	38

are placed in Group 2 because only part of the photochemical activity is restored by mild trypsin digestion. Inhibitors in Group 3 (nitrofen and trifluralin) are still inhibitory to electron transport after trypsin treatment.

The above results imply that DCMU-type inhibitors (Group 1), have one site of action on or near the "B-protein" complex that is located upon the external surface of the photosynthetic membrane. They specifically bind (non-specific binding is not taken into account) to this protein. On the other hand, ioxynil and *i*-dinoseb (Group 2) seem to affect another site of action on the O_2 evolving system. This is located presumably on the inside of the thylakoid membrane. These two inhibitors do not lose their inhibitory potency towards electron transport because a part of their activity lies in an area that is not easily accessible to trypsin. The Group 2 inhibitors also inhibit silicomolybdate-mediated O_2 evolution (data not shown). This reaction is essentially insensitive to DCMU (18) and DCMU-like inhibitors (4, 16).

The Group 3 chemicals (nitrofen and trifluralin), for which the sensitivity of electron transport is not lost by trypsin, also have a site of action different from that of DCMU-type inhibitors.

Competition Studies. By using a radiolabelled DCMU-type inhibitor, it is possible to determine if all compounds bind at the same chloroplastic site (21). ^{14}C-atrazine was added to uncoupled chloroplasts with or without other photosynthetic inhibitors. Radioactivity in the supernatant was measured after incubation and centrifugation. Values of free and bound- ^{14}C-atrazine are displayed using double reciprocal plots (Figure 3). Atrazine binds competitively with DCMU, lenacil, pyramin, propanil, RU 21731, and barban. On the contrary, nitrofen and trifluralin do not compete with atrazine for binding at the same chloroplastic site, excluding them, again, from the group of DCMU-type inhibitors. These compounds might interfere with a protein that is different from the binding site of DCMU-type chemicals. Ioxynil competes with atrazine for binding within chloroplasts, however, its affinity is much lower, relative to its inhibitory potency, than DCMU. This observation can be explained by a partitioning of the compound within the thylakoid membrane, a part only being bound to or near the B protein complex. There is no evidence for a secondary binding site at or near the oxygen evolving system, although an effect of ioxynil on the oxidizing side of PS II has been proposed herein. Other investigators have reported that *i*-dinoseb binds non-competitively with DCMU in chloroplasts (22).

Several lines of evidence suggest that *N*-phenylureas, *N*-phenylcarbamates, *N*-acylanilides, *s*-triazines, triazinones, pyridazinones, and carbamoyl 1,2,4-thiadiazolines block photosynthetic electron transport by changing the redox potential of

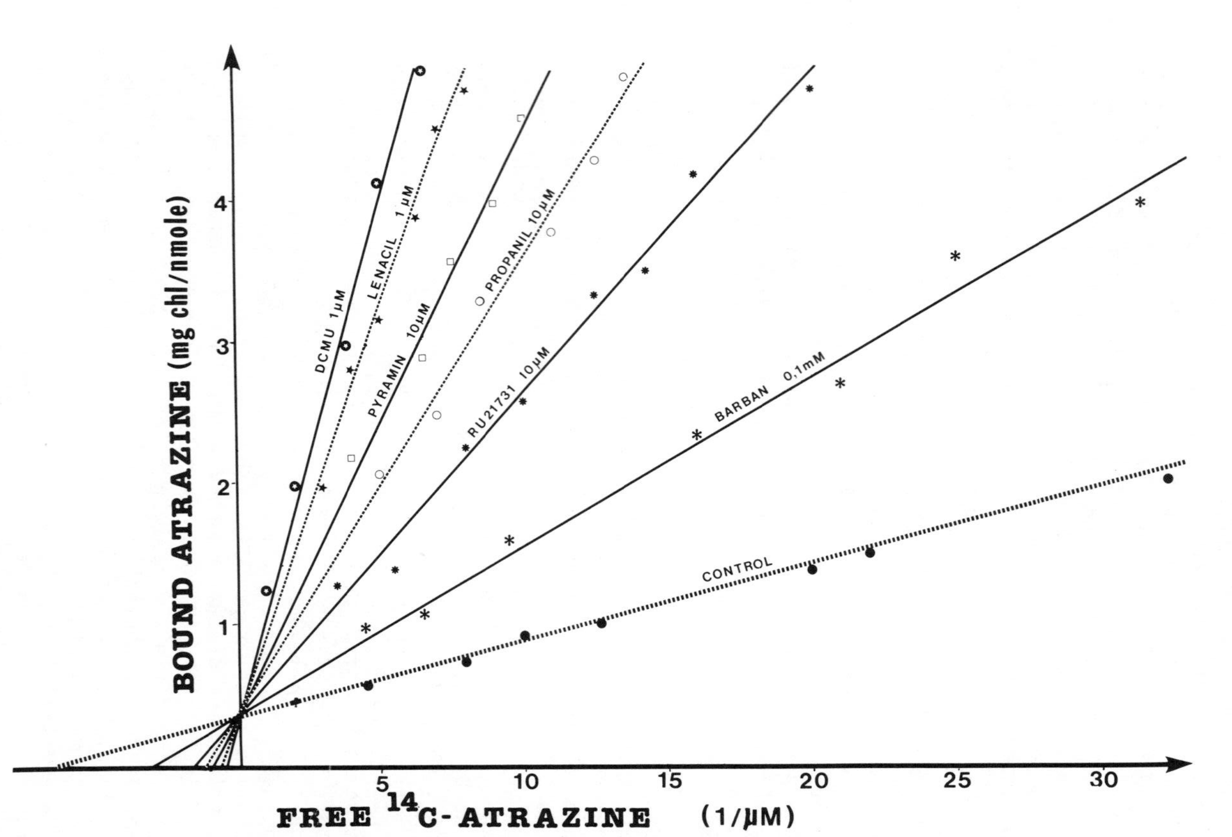

Figure 3a. Double reciprocal plot for binding of ^{14}C-atrazine in the presence of DCMU, lenacil, pyramin, propanil, RU 21731, and barban.

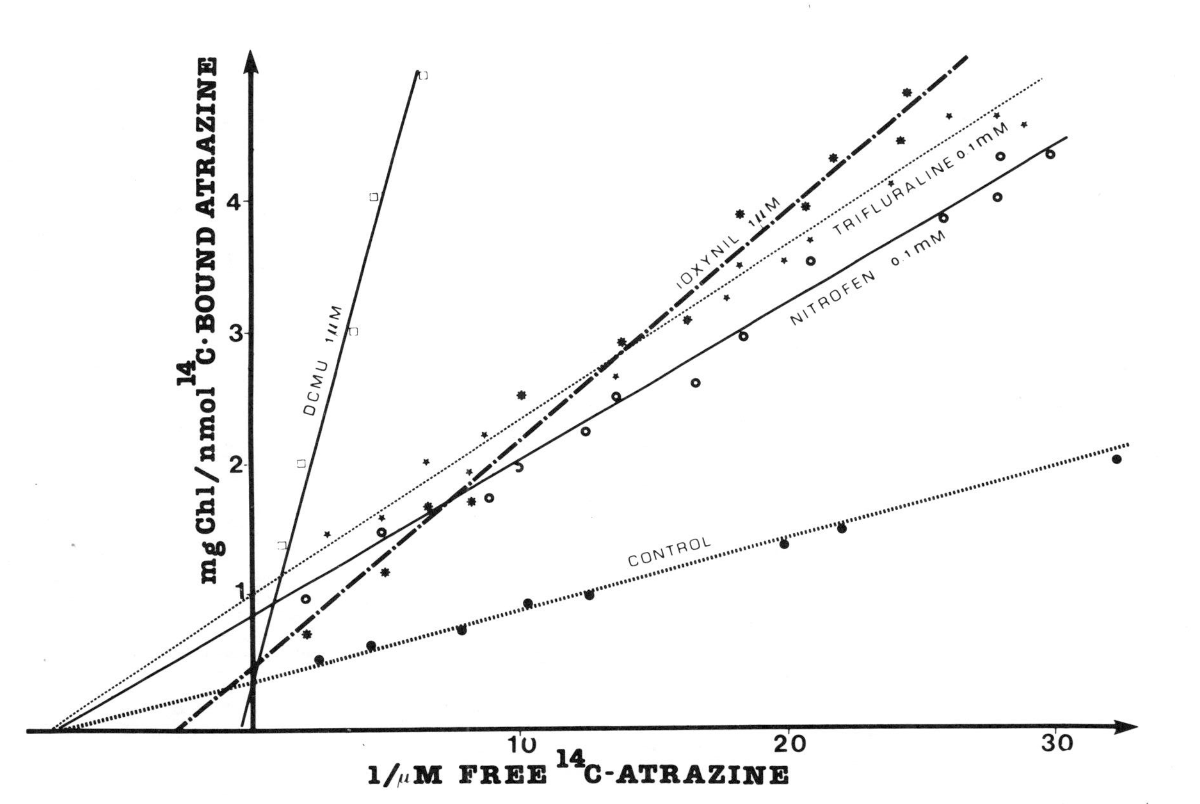

Figure 3b. Double reciprocal plot for binding of ^{14}C*-atrazine in the presence of DCMU, ioxynil, trifluraline, and nitrofen.*

the PS II secondary electron acceptor B. However, dinitroanilines and diphenylethers do not seem to act at the same site. Halogenated hydroxybenzonitriles and dinitrophenols seem to affect both sides of PS II. However, the exact type of interaction between these latter inhibitors and the common target remains to be identified.

Interaction Between Inhibitors of PS II - Catalyzed Electron Transport and a Simulated Target

Binding of photosynthetic inhibitors to the chloroplast has been shown to involve weak, noncovalent forces, suggesting that hydrogen bonds might be involved (23). A common structural feature, i.e., a -C-NH, sp_2 electron-deficient group linked to a lipophilic carrier, seems to be required for inhibitory activity (24). Actually, as it has been pointed out by several authors (see for example 25), this group might undergo hydrogen bonding with some chloroplastic active center, probably a protein. The hydrogen bonding postulate has not received experimental support. We have approached this problem using techniques derived from basic concepts of quantum chemistry.

Interactions Map with a Positive Charge: Isopotential Curves. The problem of interactions between a molecule and an ionic center can be approached by considering the event as a perturbation of the ground state of the molecule under the influence of a point charge. Using molecular orbital theory, electron motions within the nuclear field of fixed atoms can be described to represent the molecule as a wave function (Born-Oppenheimer approximation). The electronic density of each point of a molecule at its ground state arises from a knowledge of its wave function; however, studying the charge distribution around each atom is more meaningful. On the other hand, we should keep in mind that the determination of interatomic distances, and bond and torsional angles are prerequisites for using the Born-Oppenheimer approximation. To simplify calculations, internal electrons located on the nucleus were not considered and only valence electrons were taken into account. We have used, in this work, a semiempirical method, the modified CNDO/S (Complete Neglected Differential Overlap) (26). The influence of charge distribution upon the environment of a molecule can be displayed by perturbing the molecule with a point charge.

The perturbation energy is expressed as the difference of energies between the molecular-charge system and the unperturbed molecule (i.e., it represents the stabilization or destabilization energies of the whole system). In this case, electrostatic forces predominate as compared with polarization energy.

Keeping in mind that only energies that stabilize the whole system, molecule-charge-unperturbed molecule, are taken into account, the energy variations brought about by a positive point charge (a proton) are evaluated and all the points of the same stabilization energies are connected together.

Curves for the same energy level, or isopotential curves, represent an interaction map between the positive charge and the inhibitor molecule, and are displayed on a Benson-plotter. Examples of graphs obtained for a phenylurea, an acetanilide, and a *s*-triazine, are shown in Figure 4. Those for a diphenylether, a 2,4-dinitrophenol, and a triazinone are shown in Figure 5. The curves show clearly that -C=0 groups (phenylamides and triazinones) or -C=N groups (*s*-triazines), are strongly attractive to a positive charge; the attraction zone is spatially wide-spread. The repulsive zone corresponds, for each molecule, to a strong attraction area for a negative charge. These observations suggest that hydrogen bonding is possible between the corresponding attraction and repulsion zones of a target molecule, possibly a protein. For the *s*-triazines, the 2-chlorine appears to be an attraction group that actually strengthens the influence of neighboring ring-nitrogen atoms. This also suggests that each triazine molecule can undergo hydrogen bonding with two sites. The N of the alkyamino groups are repulsive to the proton, whereas the N atom in position 5 on the ring is somewhat attractive. Nitrofen, which is not a DCMU-type inhibitor, presents a different case: the molecule is somewhat twisted, so the interaction map depends upon the relative position between point charge and the nitrofen molecule. In Figure 5, the positive charge is shown being in the plane of the *p*-nitrophenyl ring and the plane of the dichlorophenyl ring is positioned at 60°. Consequently, the chlorine in position 2 does not appear to be attractive to a positive charge; however, the chlorine in position 4 is attractive. Therefore, it can be predicted that if the study was made in the plane of the chlorine-substituted ring, we would see an attraction zone around the two chlorines, as observed with DCMU or the acetanilide. Also, the attraction zone around the NO_2 group would not be displayed in the same manner. Nevertheless, the attraction properties of the two chlorines are relatively small in comparison to the NO_2 group, whereas the properties of the ether oxygen are not displayed.

The dinoseb derivative, which does not bind competitively with DCMU to chloroplasts (22), also presents a different interaction map (Figure 4): a repulsion zone appears near the 2-alkyl substituent, but because this group is rather bulky, the chances for hydrogen bond formation remain small. On the other hand, NO_2 substituents clearly show strong attraction zones, whereas the hydroxyl group is weakly attractive to a proton.

Nitrofen and the dinoseb-derivative present two cases of somewhat different properties relative to other inhibitors: it seems that the NO_2 group greatly modifies the behavior of compounds towards a proton; however, it is difficult to predict to what extent the attraction and repulsion energies located on an

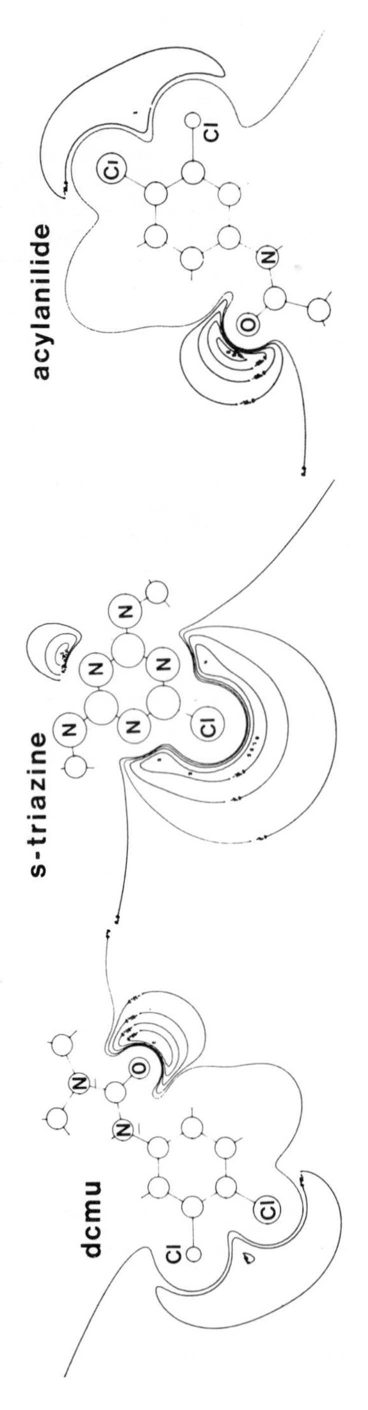

Figure 4. Benson plotter representation of interaction maps between a proton and DCMU (N-phenylureas), an s-triazine [2-chloro-4,6-bis(methylamino)-s-triazine], and an acylanilide [N-(3,4-dichlorophenyl)methylanilide].

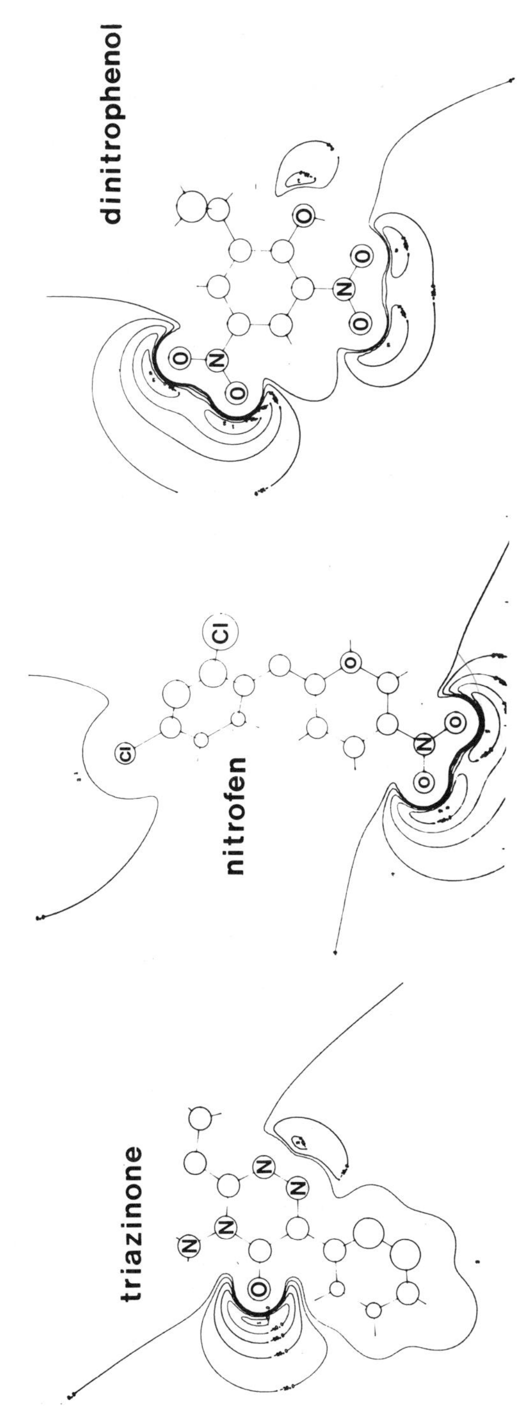

Figure 5. Interaction maps between a proton and a triazinone [4-amino-6-phenyl-3-ethyl-1,2,4-triazine-5(4H)-one], nitrofen (2,4-dichlorophenyl p*-nitrophenyl ether), and a dinitrophenol (2-isopropyl-4,6-dinitrophenol). For the latter compound, only two carbon atoms of the isopropyl group are shown.*

inhibitor molecule for a proton can represent actual hydrogen bonding with some chloroplastic site. The approach used herein involved perturbation techniques, so calculations have been carried out predominantly with long-range interactions. A proton would tend to bond preferentially with electron-rich atoms: oxygens from nitro groups are a typical case, but these examples show the limitations of the method. The chloroplastic binding site is definitely different from a proton; therefore, another approach has to be proposed for a more accurate understanding of the interactions involved.

Interaction Between Some Photosynthetic Inhibitors with a Simulated Peptidic Target. The configuration of the chloroplastic binding site, probably a 32 Kdalton protein (6), remains to be elucidated. Considering the hydrogen bonding postulate as a working hypothesis for binding to the inhibitory site, it appeared that studying a whole protein, as such, was impossible, not only because of computer limitations, but also because neither the amino acid sequence nor the conformation of the protein are known.

We previously selected and described a dipeptide (7), the conformation of which was available in the literature, that originated from quantum mechanical methods as studied by Pullman (27). The dipeptide was under the so-called H-7 form, in which the conformation of the NH-C^{α}-HR'-CO group remains sterically independent from the neighboring residues. Using the PCILO method (Perturbative Configuration Interaction using Localized Orbitals), the only possibility for the formation of one intermolecular hydrogen bond between H of the NH, O of the amide carbonyl, and other atoms of an inhibitor, occurred when the C^{α} was substituted with an aromatic residue. Instead of considering the inhibitor-dipeptide system as a supermolecule, short- and long-range intermolecular energies were evaluated by measuring mutual interactions between inhibitor and the dipeptide. Again, the entire system was considered as a perturbation, encompassing not only electrostatic force, but also polarization energy (charge distribution of the dipeptide molecule upon the inhibitor dipole and vice-versa) and dispersion energy (dipole-dipole interactions).

By moving inhibitor molecules about the dipeptide according to the above, it appears that from the energetic relations, a "preferred" position can be obtained. Using a 4014 Tektronic Computer Display Terminal, this relative position was visualized (Figure 6): the graph is simplified by showing the bonds between atoms as straight lines. This has been done for DCMU, a triazinone, a pyridazinone, and a *s*-triazine. The interatomic distance between the O (from carbonyl) and H (from NH group) of the dipeptide is 3.3 Å, corresponding closely to the interatomic distances between attractive and repulsive groups of the inhibitors (2.6 to 3.4 Å). On the other hand, the whole system

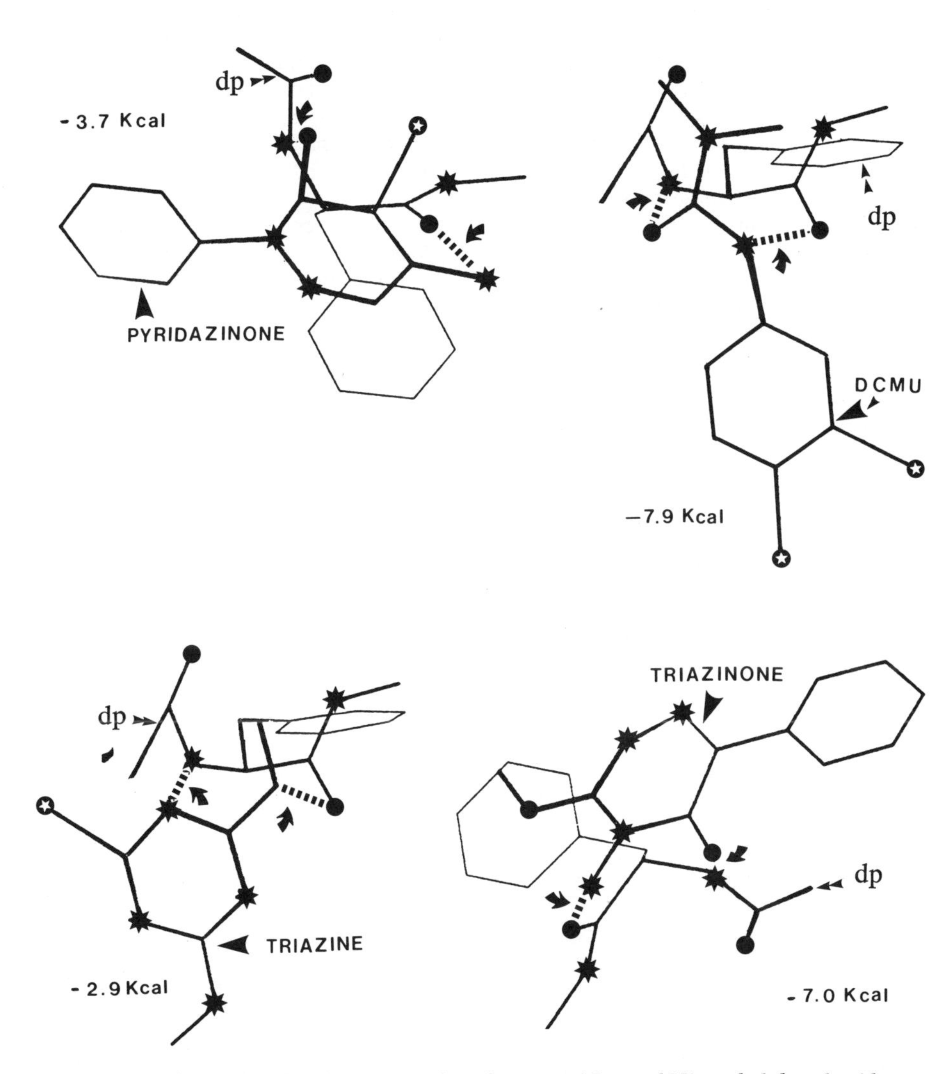

Figure 6. Intermolecular interaction between N-*acetyl*-N′-*methylphenylamide (dipeptide, dp) and photosynthesis inhibitors. Key:* ✱, *N atom;* ●, *O atom;* ✪, *Cl atom; and* ➚, *hydrogen bonds.*

(inhibitor-dipeptide) obeys a more universal law of chemical recognition. The simulated target (dipeptide) and probably the B-protein complex itself, can be considered as a receptor that orients the inhibitor molecule so that its attractive and repulsive groups form hydrogen bonds with the atoms of opposite charge on the protein. Figure 6 also shows clearly that in their preferred relative position, rings of the inhibitor molecules do not overlap the dipeptide. If the proposed model corresponds to the *in situ* situation, we can predict that the inhibitor rings present a spatially free access to any external influence within the microenvironment of the target.

From other lines of evidence that are outside the scope of this paper, we have seen that there is a chemical interaction linked to the inhibitor's ring; therefore, we assume that hydrogen bonds are a prerequisite for binding to the postulated "receptor" protein B complex, leaving free access for any interaction between the ring and some other site of the B complex, probably the plastoquinone moiety, provided the inhibitors are acting at the same site as DCMU.

Conclusion

We have accumulated evidence for a common site of action for *N*-phenylureas, *N*-acylanilides, *N*-phenylcarbamates, *s*-triazines, triazinones, pyridazinones, uracils, and carbamoyl-1,3,4-thiadiazolines. Halogenated hydroxybenzonitriles and 2,4-dinitrophenols interfere with electron transport at two sites, one on the donor side of PS II, the other at the level of the PS II secondary electron acceptor B. However, *i*-dinoseb does not seem to act exactly like other DCMU-type inhibitors. From molecular orbital data, it appears that DCMU-type inhibitors probably bind through hydrogen bonds to a chloroplastic protein, in an area presumably close to an aromatic amino acid residue. We recognize that simulating the binding site of the B-protein complex with a dipeptide is a crude approximation. Additional studies are underway to check the binding properties of ioxynil and *i*-dinoseb.

We assume the binding of photosynthetic inhibitors is a prerequisite for activity, which is in turn not necessarily directly related to a common structural feature, a-N-C̈, sp_2 electron - deficient group. However, it is difficult to separate binding from inhibition activity towards electron transport. Intermolecular energies of the inhibitor-dipeptide system seem to rank in the same order as binding constants of the inhibitors with the chloroplast (see for example ref. 21).

Diphenylethers and dinitroanilines show both a different site of action relative to DCMU, and an interaction map that is difficult to identify with DCMU-type inhibitors. They may bind to another protein on the thylakoid membrane.

Literature Cited

1. Moreland, D. E. Ann. Rev. Plant Physiol. 1980, 31, 597-638.
2. Velthuys, B. R.; Amsez, J. Biochim. Biophys. Acta 1974, 333, 85.
3. Van Assche, C. J. "Advances in Pesticide Science"; Geissbüehler, H., Ed.; Part 3; Pergamon Press: Oxford, 1979; p. 494-8.
4. Van Assche, C. J. Thèse Doctorat d'Etat, 1981.
5. Pulles, M. P. J.; Van Gorkom, H. J.; Willemsen, J. C. Biochim. Biophys. Acta 1976, 449, 536-40.
6. Steinback, K. E.; Pfister, K.; Arntzen, C. J.; Chapter 3 of this book.
7. Carles, P. M.; Van Assche, C. J. Proc. 5th Int. Congr. Photosynth.: Halkidiki, Greece, 1981; in press.
8. Murata, N.; Nishimura, M.; Takamiya, A. Biochim. Biophys. Acta 1966, 120, 23-33.
9. Forbush, B; Kok, B. Biochim. Biophys. Acta 1968, 162, 243-53.
10. Etienne, A. L.; Lemasson, C.; Lavorel, J. Biochim. Biophys. Acta 1974, 333, 288-300.
11. Munday, J. L.; Govindjee Biophys. J. 1969, 9, 1-21.
12. Wollman, F. A.; Thorez, D. S. C. R. Acad. Sc. Paris 1976, t. 285, série D, 1345-8.
13. Wollman, F. A. Biochim. Biophys. Acta 1978, 503, 263-73.
14. Lavorel, J. "Bioenergetics of Photosynthesis"; Govindjee, Ed.; Academic Press: New York, 1975; pp. 223-317.
15. Etienne, A. L.; Lavorel, J. FEBS Lett. 1979, (57) 3, 276-9.
16. Van Assche, C. J. Proc. 5th Int. Congr. Photosynth.: Halkidiki, Greece, 1981; in press.
17. Lavorel, J. Biochim. Biophys. Acta 1980, 590, 385-99.
18. Giaquinta, R. T.; Dilley, R. A. Biochim. Biophys. Acta 1975, 387, 288-305.
19. Renger, G. Biochim. Biophys. Acta 1976, 440, 287-300.
20. Pfister, K.; Arntzen, C. J. Z. Naturforsch. 1979, 34c, 996-1009.
21. Tischer, W.; Strotmann, H. Biochim. Biophys. Acta 1977, 460, 113-25.
22. Ottmeier, W.; Masson, K. Pestic. Biochem. Physiol. 1980, 14 (1), 86-97.
23. Moreland, D. E. "Progress in Photosynthesis Research"; Vol. III; Metzner, H., Ed.; H. Laupp, Jr.: Tübingen, Germany; 1969; p. 1693-1711.
24. Hansch, C. "Progress in Photosynthesis Research"; Vol. III; Metzner, H., Ed.; H. Laupp, Jr.: Tübingen, Germany; 1969; p. 1685-92.
25. Camper, N. D.; Moreland, D. E. Biochim. Biophys. Acta 1965, 94, 383-93.
26. Rajzmann, M.; Francois, P.; Carles, P. M. J. Chim. Phys. 1979, 76, 328.
27. Pullman, B. "Aspects de la Chimie Quantique Contemporaine"; Colloques Internationaux du C.N.R.S. n° 195, Menton 1970, p. 261.

RECEIVED September 14, 1981.

2

Binding Sites Associated with Inhibition of Photosystem II

LESTER L. SHIPMAN[1]

Argonne National Laboratory, Argonne, IL 60439

A variety of experimental and theoretical evidence has been integrated into coherent molecular mechanisms for the action of photosystem II herbicides. Photosystem II herbicides act by inhibiting electron transfers between the first and second plastoquinones on the reducing side of photosystem II. Each herbicide molecule must have a flat polar component with hydrophobic substituents to be active. The hydrophobic substituents serve to partition the molecule into lipid regions of the cell and to fit the hydrophobic region of the herbicide binding site. The flat polar portion of the herbicide is used for electrostatic binding to the polar region of the herbicide binding site. Theoretical calculations have been carried out to investigate the binding of herbicides to model proteinaceous binding sites.

Many commercially and scientifically important herbicides act directly on components in the photosynthetic membrane of green plants. Usually, these herbicides act by inhibiting electron transport in the photosynthetic electron transport chain. In the photosynthetic membrane, electrons are ultimately removed from water (evolving molecular oxygen as a byproduct) and are given to $NADP^+$. For reviews on the action of herbicides, see discussions by Ashton and Crafts (1), Corbett (2), Audus (3), and Moreland (4). Of the photosynthetic herbicides presently in commercial use, the majority act at a special site on the reducing side of photosystem II (PS II) within the photosynthetic membrane. This particular class of herbicides is called "PS II herbicides". There

[1] Current address: E. I. du Pont de Nemours & Co., Inc., Central Research and Development Dept., Experimental Station, Wilmington, DE 19898.

0097-6156/82/0181-0023$05.00/0

are two plastoquinone (PQ) molecules (see Figure 1) embedded in the PS II complex; electrons are transferred sequentially from the first (Q) to the second (B) plastoquinone. The PS II herbicides bind noncovalently to a special high-affinity site on PS II, inhibit electron transfer from Q^- to B, and effectively shut down electron transport through PS II. A wide variety of chemical structures are found amoung the PS II herbicides including phenylureas, amides, s-triazines, triazinones, nitrophenols, carbamates, bis-carbamates, pyridazinones, hydroxybenzonitriles, and uracils. The deleterious effects of PS II herbicides on plant photosynthesis have been probed primarily through their effects on chlorophyll fluorescence coming from PS II as well as their effects upon electron transport through PS II (5-35).

Recently, advanced theoretical chemical techniques including *ab initio* molecular quantum mechanics have been used to investigate the PS II herbicides (36). These theoretical methods have been used to seek similarities as well as differences between the properties of PS II herbicides known to compete for the same or nearby binding sites. In addition, theoretical techniques have been used to study the energetics and geometries for the interaction of PS II herbicides with local protein structures.

Several recent experimental papers have reported exciting new results on the protein component of PS II that binds the PS II herbicides (37-43). There is general agreement among these studies that the molecular weight of the herbicide-binding proteins are in the range 30-40 kDaltons.

The present paper is a discussion of the photosystem II herbicides and their mechanisms of action. Among the topics covered are the green plant photosystems, photochemistry and electron transfers within photosystem II, requirements for herbicidal activity, mechanisms of action, herbicide selectivity and resistance, herbicide-binding proteins, and theoretical studies of herbicide-binding site interactions.

Green Plant Photosystems

The green plant photosynthetic membrane has two types of photosystems, photosystem I (PS I) and PS II. For a review of the structure and function of the photosynthetic membrane, see refs. 44,45,46. For a review of PS II in particular, see ref. 47. Photosystem II uses the energy from absorbed photons to remove electrons from water on the inside of the photosynthetic membrane and give them to plastoquinone. PS I removes electrons from plastohydroquinone and donates them to $NADP^+$. Thus, overall, water is oxidized (O_2 is evolved) and $NADP^+$ is reduced by the cooperative action of PS I and PS II. Each photosystem is a complex of protein with many pigments such as chlorophylls (Chls) and carotenoids as well as electron transfer components such as quinones, cytochromes, and iron-sulfur centers. The most

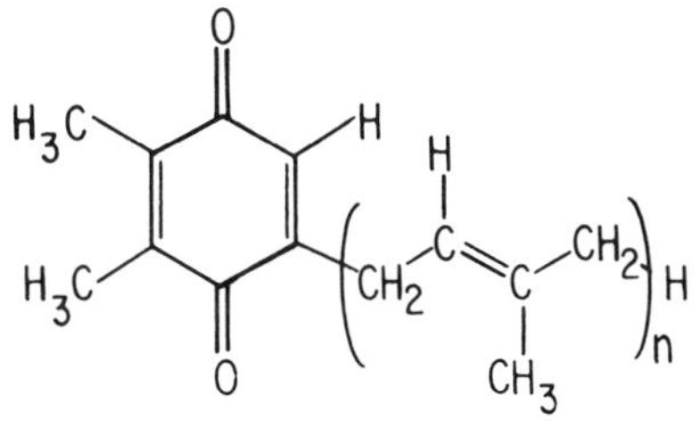

Figure 1. Chemical structure for plastoquinone. The value of n in the polyisoprene tail is 9 for the most abundant plastoquinone in green plants, plastoquinone A.

abundant pigment molecule in PS I and PS II as well as the molecule that functions as the primary electron donor is chlorophyll a.

Photochemistry and Electron Transfers Within Photosystem II

Most of the proteins of PS II are embedded within the photosynthetic membrane although portions are exposed to the aqueous media on the inside and on the outside of the membrane. A fully developed PS II is composed of hundreds of Chl a and Chl b molecules, carotenoids, plastoquinones (Figure 1), α-tocopheryl quinone or α-tocopherol, cytochrome b-559, the Mn-protein responsible for O_2 evolution, and other electron transport agents. The diameter of a fully-developed PS II has been estimated at 160 Å from electron micrographs of freeze-fractured photosynthetic membranes (46). The bulk of the chlorophyll molecules in PS II have only an antenna function, i.e., they absorb photons (reaction 1) and transfer the resultant electronic

Chl(antenna) + hν → Chl(antenna)* + heat (reaction 1)

excitation quickly to the primary electron donor, P680, (reaction 2) in the PS II reaction center where the primary

Chl(antenna)* P680 → Chl(antenna) P680* (reaction 2)

electron transfer event takes place. The electronic excitation moves through the antenna pigments in the form of singlet excitons (48,49).

The electron transfer chain on the reducing side of PS II consists of at least P680, a pheophytin molecule (Pheo), and two plastoquinone molecules (Figure 1) Q and B. The ground state of the reaction center is state 1. When a singlet exciton reaches

P680 Pheo Q B (state 1)

the reaction center, P680 becomes electronically excited to P680* and state 2 is populated. An electron is then ejected from P680*

P680* Pheo Q B (state 2)

to a nearby Pheo molecule (50-55) with a possible intermediate transfer across a Chl a molecule and state 3 is populated.

$P680^+$ $Pheo^-$ Q B (state 3)

Evidence for the possible participation of an intermediate Chl a molecule comes from a recent ESR study (53) and an analogy with the properties of the corresponding photosystem (56) in purple photosynthetic bacteria. When the electron transfer chain on the reducing side of PS II is blocked, the radical pair corresponding to ($P680^+$ $Pheo^-$) must collapse back to a lower triplet state or to the ground singlet state. This collapsed triplet state is highly spin-polarized (53), i.e., the middle triplet sublevel ($m_s = 0$) is preferentially populated. The spin polarization pattern indicates that electronic exchange between the two radicals in the radical pair is small, which in turn suggests that P680 and Pheo are not in contact. Further study will be necessary to prove or disprove the participation of an intermediary Chl a. A similar spin-polarized triplet has been observed previously in purple photosynthetic bacteria (57,58). An electron transfer from $Pheo^-$ to Q populates state 4 from

$P680^+$ Pheo Q^- B (state 4)

state 3. At this point $P680^+$ is reduced via an electron transfer from electron donor Z, an intermediary between P680 and the Mn-protein responsible for oxygen evolution. Upon reduction of $P680^+$ and transfer of an electron from Q^- to B, state 5 is

P680 Pheo Q B^- (state 5)

populated. If another photon is not absorbed by PS II in state 5, state 5 is stable for many seconds (59,60). Interestingly, optical spectra indicate that reduced B is in the form B^-, not BH (60-63); the local protein environment around B must stabilize the anion form over the neutral form and/or provide a barrier for the uptake of protons onto B^-. Absorption of a second photon by PS II (reaction 1) and migration of the resultant singlet exciton to P680 (reaction 2) leads to the population of state 6. State 7 is populated from state 6 by the

P680* Pheo Q B^- (state 6)

$P680^+$ $Pheo^-$ Q B^- (state 7)

ejection of an electron from P680* to a nearby Pheo (again with a possible intermediate transfer across a Chl a molecule). State 8 is populated from state 7 by the transfer of an electron

$P680^+$ Pheo Q^- B^- (state 8)

from $Pheo^-$ to Q. $P680^+$ is then reduced by Z to populate state 9.

P680 Pheo Q^- B^- (state 9)

Electronic absorption spectra of the plastoquinone show that the two electrons plus two protons rapidly end up on the same plastoquinone molecule, PQH_2. State 10 may be transiently

P680 Pheo Q B^{2-} (state 10)

populated before the protons arrive on plastoquinone B. The protons do not come directly from the external aqueous media, they must pass through a barrier, probably a protein. Proton uptake from the external aqueous phase has been measured optically by use of pH-sensitive dyes; this proton uptake from aqueous media is almost two orders of magnitude slower than electron transport to PQ (64). The detailed mechanism and kinetics for the two-electron, two-protein reduction of PQ to PQH_2 have not, as yet, been worked out and further studies are indicated. From state 10, the reducing side of PS II recycles to state 1. The oxidizing side of PS II, however, has a four photon cycle with an O_2 molecule evolved on every cycle. The four photon cycle of the oxidizing side is to be contrasted with the two-photon cycle on the reducing side.

Requirements for Herbicidal Activity

As mentioned in the introduction, molecules from a number of different classes of herbicides are PS II inhibitors. A theoretical analysis has been carried out to isolate the molecular properties shared by all active PS II herbicides (36). When the results of this theoretical analysis were combined with experimental data on transport and metabolism, the following list of required properties was developed for active PS II herbicides.

1. The herbicide molecule must have a flat polar component approximately the size of a phenyl ring.
2. The flat polar component must have hydrophobic substituents such that it is partitioned strongly into lipid.
3. The herbicide molecule must be readily absorbed and transported through many partitions between the point of contact (roots or foliage) and the site of action (chloroplast).
4. The metabolism of the herbicide must be very slow because the binding to the active site is noncovalent and reversible, and the herbicide must be bound long enough for the deleterious action of light and oxygen to take effect.

Mechanisms of Action

As previously discussed, the PS II herbicides act by binding noncovalently to a special proteinaceous site (or overlapping sites) on the reducing side of PS II and inhibit electron transfer from Q^- to B. Because the available experimental evidence shows only that the electron stays on Q^- and does not directly probe B, at least two mechanisms are consistent with available data.

Mechanism 1. Inhibitory herbicides displace plastoquinone B from its proteinaceous binding site on the reducing site of PS II.

Evidence in support of mechanism 1 is (i) the displacement of ubiquinone by the inhibitor orthophenanthroline in the analogous photosystem found in purple photosynthetic bacteria (56) and (ii) the similarity in size and shape between the flat polar component of the PS II herbicides and the quinone head of plastoquinone. Mechanism 1 can be studied directly through competitive displacement reactions between PS II herbicides and PQ.

Mechanism 2. Herbicide binding shifts the redox potential for B to more negative values, i.e., when the herbicide is bound, B is harder to reduce.

This shift in redox potential could arise from (i) a direct coulombic interaction between B^- and the polar component of the bound herbicide and/or (ii) inhibition of the relaxation of the B-binding protein during the $B \rightarrow B^-$ transformation, and/or (iii) inhibition of proton uptake by the protein on the reducing side of PS II during the $B \rightarrow B^-$ transformation.

Further experimental work is needed to choose between mechanisms 1 and 2.

Blockage of electron transport on the reducing side of PS II is just the first in a series of steps that ultimately leads to plant death. Much of the energy from absorbed photons that is normally directed into electron transport is redirected into fluorescence and triplet formation when the herbicides are bound. The triplet states are of special interest because of the destructive interaction between excited chlorophyll and molecular oxygen through the following four step mechanism, where

$${}^{S_0}Chl + h\nu \rightarrow {}^{S_1}Chl \qquad \text{(reaction 3)}$$

$${}^{S_1}Chl \rightarrow {}^{T_0}Chl + heat \qquad \text{(reaction 4)}$$

$${}^{T_0}Chl + {}^{T_0}O_2 \rightarrow {}^{S_0}Chl + {}^{S_0}O_2 + heat \qquad \text{(reaction 5)}$$

$${}^{S_0}Chl + {}^{S_0}O_2 \rightarrow \text{oxidized chlorophyll derivatives} \qquad \text{(reaction 6)}$$

${}^{S_0}Chl$, ${}^{S_1}Chl$, ${}^{T_0}Chl$, ${}^{T_0}O_2$, and ${}^{S_0}O_2$ are the ground single state of Chl, the lowest excited singlet state of Chl, the lowest triplet state of Chl, and ground triplet state of O_2, and the lowest singlet state of O_2, respectively. Fortunately, the chlorophylls are somewhat protected in vivo by the rapid transfer of triplet excitons from chlorophylls to carotenoids (65) via reaction 7. This protection of Chl is paid for by the occasional

$${}^{T_0}Chl + {}^{S_0}Car \rightarrow {}^{S_0}Chl + {}^{T_0}Car + heat \qquad \text{(reaction 7)}$$

photodestruction of carotenoids. Consistent with this picture is the observation that carotenoid pigments tend to be bleached before chlorophyll pigments in the presence of PS II herbicides, oxygen, and light (66).

Although the mechanism involving stoppage of electron transfer in PS II followed by photodestruction is currently thought to be the major mode of action for the PS II herbicides, it should be kept in mind that many herbicides have multiple modes of action. Exogenous molecules such as herbicides can and do induce a variety of physiological and biochemical changes in the plant. Of particular interest are the "inhibitory uncouplers" (4) (e.g., nitrophenols) which not only bind specifically to PS II and inhibit electron transfer, but also interact nonspecifically with membranes and break down the pH gradient established across photosynthetic and mitochondrial membranes.

Herbicide Selectivity and Resistance

The selectivity of a particular herbicide toward a specific plant is based upon differences between plants for a number of factors including, but not limited to, the following.

1. Rate of uptake.
2. Transport rates and partitioning.
3. Metabolism rate.
4. Binding site affinity.

For the particular case of triazine-resistant weed biotypes found in areas of the world where there has been frequent use of triazine herbicides, the resistance has been traced to a lowered binding affinity at the PS II herbicide binding site (17,19,25,26).

Herbicide-Binding Proteins

Proteins provide the framework upon which the light-absorbing, exciton-transferring, and electron-transferring components are arranged. A random collection of manganese ions, chlorophylls, pheophytin, and plastoquinones would not have the properties that we associate with PS II. The proteins hold the electron transfer components in proper alignment and separation with respect to each other so that a high quantum yield for forward electron transfer is obtained while giving a low quantum yield for the cation hole-electron recombination.

Trypsin is a water-soluble protease that cleaves polypeptide chains preferentially at arginine and lysine residues. Thus, when trypsin is applied to broken chloroplasts or PS II particles the surface exposed proteins are digested first and the fully buried proteins are initially left alone. A number of studies (10,22,29,30,38,41,42,43,67) have been conducted to determine the effects of digestion of the surface-exposed polypeptides on the function of the reducing side of PS II. Trypsin treatment impairs the ability of PS II herbicides to inhibit the reduction of water-soluble electron acceptors such as ferricyanide by PS II. Extensive digestion of the photosynthetic membrane eventually destroys the herbicide binding site(s). Trypsin digestion of PS II particles leads to the loss of polypeptides of molecular weight 27 and 32 kDaltons (22); this is consistent with the molecular weights of the herbicide binding proteins as determined through photoaffinity labeling experiments.

PS II herbicides that are converted to photoaffinity labels have proved quite useful for the identification of the proteins that participate in herbicide binding. In a typical experiment, the photoaffinity label is bound noncovalently to the PS II herbicide binding site in the dark. The system is then irradiated with UV light and after irradiation the photoaffinity label is bound covalently at or near its original binding site. In particular, ^{14}C-labelled azidoatrazine (4-azido-2-isopropylamino-6-ethylamino-*s*-triazine) has been shown to bind to a polypeptide of molecular weight 32 kDaltons (37,38,43). A ^{3}H-labelled azido derivative of dinoseb (2-azido-4-nitro-6-isobutylphenol) was shown to bind to polypeptides in the 30-40 kDalton range (40). It may be concluded that the polypeptides in PS II that are responsible for binding the PS II herbicides have molecular weights in the 30-40 kDalton range; one of them having a molecular weight of 32 kDaltons.

Theoretical Studies of Herbicide-Binding Site Interactions

In a recent theoretical study, Shipman (36) proposed several specific models for the herbicide binding site on PS II. It was proposed that the PS II herbicides bind electrostatically at or near a protein salt bridge or the terminus of an alpha helix. The salt bridge could be argininium aspartate, argininium glutamate, lysinium asparate, lysinium glutamate, argininium lysine-carbamate, or lysinium lysine-carbamate. Both the salt bridges and the terminus of an alpha helix generate very strong local electric fields with which the polar component of the herbicide molecules could bind electrostatically, perhaps with additional stabilization from hydrogen bonds. The strong local electric field from the protein, if suitably oriented, could stabilize B^- relative to B; this would explain, in part, the stability of B^- for many seconds. Adding to the binding site model the requirement that it be buried in a hydrophobic environment away from the aqueous phase, the dual requirements of lipophilicity and polarity for the PS II herbicides are explained.

Pi-pi complex between a model for a polar component of a herbicide (uracil) and a model for a salt binding site (guanidinium bicarbonate) is shown in Figure 2. The geometry was computed from *ab initio* calculations (36). It should be noted that in the lowest energy geometry shown in Figure 2, the dipole moment of uracil is antiparallel to the dipole moment of the salt.

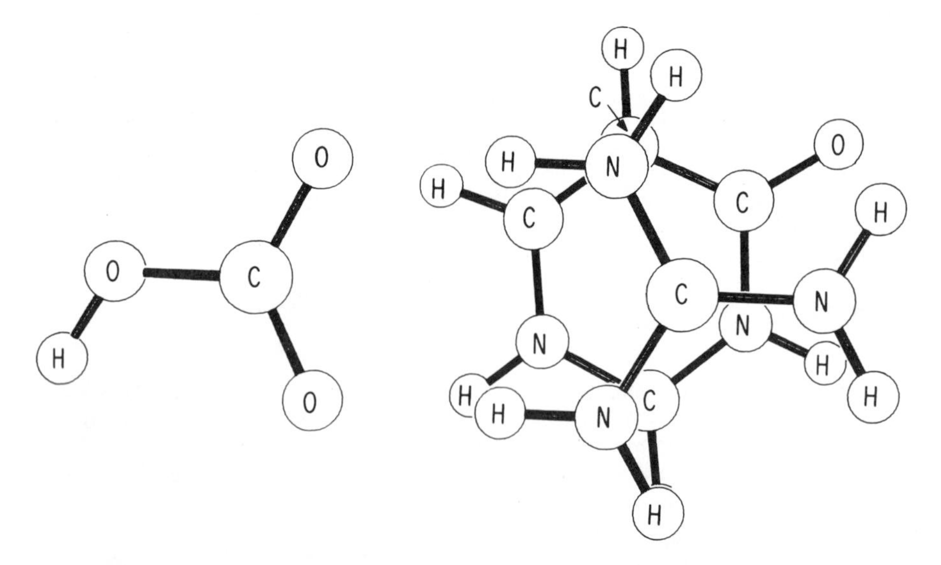

Figure 2. Lowest energy geometry for the π–π complex between a model for the polar component of a PS II herbicide (uracil) and a salt model (guanidinium bicarbonate). The circles around the atoms were drawn at 20% of their van der Waals radii (36).

Carles and Van Assche (68,69) have proposed a dipeptide model for the herbicide binding site and have used the CNDO/s semi-empirical quantum mechanical method to study herbicide-dipeptide interactions. Two hydrogen bonds are important for the energetics of herbicide binding to the dipeptides. Coulombic isopotential maps were used as a guide to the way the herbicide molecules might interact with protein.

Acknowledgment

Work supported by the Division of Chemical Sciences, Office of Basic Energy Sciences, U.S. Department of Energy, under Contract W-31-109-Eng-38.

Literature Cited

1. Ashton, F. M.; Crafts, A. S. "Mode of Action of Herbicides"; J. Wiley & Sons: New York, 1973, 504 pp.
2. Corbett, J. R. "The Biochemical Mode of Action of Pesticides"; Academic Press: New York, 1974, 330 pp.
3. Audus, L. J. "Herbicides"; Vol. 1; Academic Press: New York, 1976, 608 pp.
4. Moreland, D. E. Ann. Rev. Plant Physiol. 1980, 31, 597.
5. Hansch, N. C. in "Progress in Photosynthesis Research"; Metzner, H. Ed.; H. Loupp, Jr.: Tübingen. 1969; p. 1685.
6. Moreland, D. E. in "Progress in Photosynthesis Research"; Metzner, H. Ed.; H. Loupp, Jr.: Tübingen, 1969; p. 1693.
7. Trebst, A.; Harth, E. Z. Naturforsch. 1974, 29C, 232.
8. Regitz, G.; Ohad, I. J. Biol. Chem. 1976, 251, 247.
9. Renger, G. FEBS Lett. 1976, 69, 225.
10. Renger, G. Biochim. Biophys. Acta 1976, 440, 287.
11. Renger, G.; Erixon, K.; Doring, G.; Wolff, C. Biochim. Biophys. Acta 1976, 440, 278.
12. Trebst, A.; Reimer, S.; Dallacker, F. Plant Sci. Lett. 1976, 6, 21.
13. Izawa, S. in "Encyclopedia of Plant Physiology"; New Series, Vol. 5; Trebst, A.; Avron, M. Ed.; Springer: Berlin, 1977; p. 266.
14. Renger, G. in "Bioenergetics of Membranes"; Packer, L. Ed.; Biomedical Press, Elsevier: Amsterdam, 1977; p. 339.
15. Tischer, W.; Strotmann, H. Biochim. Biophys. Acta 1977, 460, 113.
16. Draber, W.; Fedtke, C. in "Advances in Pesticide Science"; Geissbühler, H., Ed.; Pergamon: Oxford, 1979; p. 475.
17. Machado, V. S.; Arntzen, C. J.; Bandeen, J. D.; Stephensen, G. R. Weed Sci. 1978, 26, 318.
18. Van Rensen, J. J. S.; Wong, D.; Govindjee Z. Naturforsch. 1979, 34C, 951.

19. Arntzen, C. J.; Ditto, C. L.; Brewer, P. E. Proc. Natl. Acad. Sci. U.S.A. 1979, 76, 278.
20. Brewer, P. E.; Arntzen, C. J.; Sliffe, F. W. Weed Sci. 1979, 27, 300.
21. Brugnoni. G. P.; Moser, P.; Trebst, A. Z. Naturforsch. 1979, 34C, 1028.
22. Croze, E.; Kelly, M.; Horton, P. FEBS Lett. 1979, 103, 22.
23. Govindjee; Jursinic, P. A. Photochem. Photobiol. Rev. 1979, 4, 125.
24. Pfister, K.; Radosevich, S. R.; Arntzen, C. J. Plant Physiol. 1979, 64, 995.
25. Pfister, K.; Arntzen, C. J. Z. Naturforsch. 1979, 34C, 996.
26. Radosevich. S. R.; Steinback, K. E.; Arntzen, C. J. Weed Sci. 1979, 27, 216.
27. Reimer, S.; Link, K.; Trebst, A. Z. Naturforsch. 1979, 34C, 419.
28. Renger, G.; Tiemann, R. Biochim Biophys. Acta 1979, 545, 316.
29. Renger, G. Z. Naturforsch. 1979, 34C, 1010.
30. Tischer, W.; Strotmann, H. Z. Naturforsch. 1979, 34C, 992.
31. Trebst, A. Z. Naturforsch. 1979, 34C, 986.
32. Urbach, W.; Lurz, G.; Hartmeyer, H.; Urbach, D. Z. Naturforsch. 1979, 34C, 951.
33. Van Rensen, J. J. S.; Hobe, J. H. Z. Naturforsch. 1979, 34C, 1021.
34. Wright, K.; Corbett, J. R. Z. Naturforsch. 1979, 34C, 966.
35. Moreland, D. E. Ann. Rev. Plant Physiol. 1980, 31, 597.
36. Shipman, L. L. J. Theoret. Biol. 1981, 90, 123.
37. Gardner, G. Science 1981, 211, 937.
38. Pfister, K.; Steinback, K. E.; Arntzen, C. J. in Proc. 5th Int. Congr. Photosynth., Halkidiki, Greece, 1981; in press.
39. Oettmeier, W.; Johanningmeier, U. in "First European Bioenergetics Conference"; Pàtron Editore: Bologna, 1980; p. 57.
40. Oettmeier, W.; Masson, K.; Johanningmeier, U. FEBS Lett. 1980, 118, 267.
41. Steinback, K. E.; Pfister, K.; Arntzen, C. J. Z. Naturforsch. 1981, 36C, 98.
42. Oettmeier, W.; Masson, K. Pesticide Biochem. Physiol. 1980, 14, 86.
43. Mullet, J. E.; Arntzen, C. J. Biochim. Biophys. Acta 1981, 635, 236.
44. Staehelin. L. A.; Armond, P. A.; Miller, K. R. in "Chlorophyll-Proteins, Reaction Centers, and Photosynthetic Membranes"; Brookhaven Symposium in Biology No. 28; Brookhaven National Laboratory: Upton, Long Island, New York, 1976; p. 278.
45. Arntzen, C. J.; Armond, P. A.; Briantais, J.-M.; Burke, J. J.; Novitzky, W. P. in "Chlorophyll-Proteins, Reaction Centers, and Photosynthetic Membranes"; Brookhaven Symposium in

Biology No. 28; Brookhaven National Laboratory: Upton, Long Island, New York, 1976; p. 316.
46. Staehelin, L. A.; Arntzen, C. J. in "Chlorophyll Organization and Energy Transfer in Photosynthesis"; Ciba Foundation Symposium 61; Excerpta Medica: New York, 1979; p. 147.
47. Amesz, J.; Duysens, L. N. M. in "Primary Processes of Photosynthesis"; Topics in Photosynthesis, Vol. 2; Barber, J. Ed.; Elsevier: New York, 1977; p. 149.
48. Knox, R. S. in "Bioenergetics of Photosynthesis"; Govindjee Ed.; Academic Press: New York, 1975; p. 183.
49. Shipman, L. L. Photochem. Photobiol. 1980, 31, 157.
50. Klimov, V. V.; Dolan, E.; Ke, B. FEBS Lett. 1980, 112, 97.
51. Klimov, V. V.; Dolan, E.; Ke, B. Proc. Natl. Acad. Sci. U.S.A. 1980, 77, 7227.
52. Shuvalov, V. A.; Klimov, V. V.; Dolan, E.; Parson, W. W.; Ke, B. FEBS Lett. 1980, 118, 279.
53. Rutherford, A. W.; Paterson, D. R.; Mullet, J. E. Biochim. Biophys. Acta 1981, 635, 205.
54. Fajer, J.; Davis, M. S.; Forman, A.; Klimov, V. V.; Dolan, E.; Ke, B. J. Am. Chem. Soc. 1980, 102, 7143.
55. Rutherford, A. W.; Mullet, J. E.; Crofts, A. R. FEBS Lett. 1981, 123, 235.
56. "The Photosynthetic Bacteria"; Clayton, R. U.; Sistrom, W. R. Ed.; Plenum Press: New York, 1978.
57. Dutton, P. L.; Leigh, J. S.; Seibert, M. Biochem. Biophys. Res. Commun. 1971, 40, 406.
58. Thurnauer, M. C.; Katz, J. J.; Norris, J. R. Proc. Natl. Acad. Sci. U.S.A. 1975, 72, 3270.
59. Velthuys, B. R.; Amesz, J. Biochim. Biophys. Acta 1974, 333, 85.
60. Pulles, M. P. J.; van Gorkom, H. J.; Willemsen, J. G. Biochim. Biophys. Acta 1976, 449, 539.
61. Stiehl, H. H.; Witt, H. T. Z. Naturforsch. 1968, B23, 220.
62. Witt, K. FEBS Lett. 1973, 38, 116.
63. Van Gorkom, H. J. Biochim. Biophys. Acta 1974, 347, 439.
64. Ausländer, W.; Junge, W. Biochim. Biophys. Acta 1974, 357, 285.
65. Kramer, H.; Mathis, P. Biochim. Biophys. Acta 1980, 593, 319.
66. Pallett, K. E.; Dodge, A. D. Pestic. Sci. 1979, 10, 216.
67. Böger, P.; Kunert, K.-J. Z. Naturforsch. 1979, 34C, 1015.
68. Carles, P. M.; Van Assche, C. J. in Proc. 5th Int. Congr. Photosynth., Halkidiki, Greece, 1981; in press.
69. Van Assche, C. J.; Carles, P. M., Chapter 1 in this book.

RECEIVED October 6, 1981.

3

Identification of the Receptor Site for Triazine Herbicides in Chloroplast Thylakoid Membranes

KATHERINE E. STEINBACK, KLAUS PFISTER[1] and CHARLES J. ARNTZEN

Michigan State University, MSU–DOE Plant Research Laboratory,
East Lansing, MI 48824

A broad range of inhibitors of photosynthesis act by blocking the same step in electron transport in chloroplast membranes. Inhibition occurs at the level of a protein-bound plastoquinone (B) that functions as the second stable electron acceptor for photosystem II (PS II). Studies based on the use of the proteolytic enzyme, trypsin, indicate that the receptor site for PS II inhibitors is a protein of the structural PS II complex. Using a photoaffinity ^{14}C-labeled derivative of atrazine, we have identified the specific receptor polypeptide for this inhibitor of PS II function. Analysis of membrane polypeptides by polyacrylamide gel electrophoresis and fluorographic techniques have shown that the photoaffinity triazine covalently binds to a polypeptide of 32-34 kilodaltons. We have further shown that this polypeptide is surface exposed; trypsin treatment of thylakoid membranes results in the stepwise alteration of the peptide to a 16 kilodalton species. The site for covalent attachment of the photoaffinity probe is located on the intrinsic 16 kilodalton fragment. Consequently, the binding site appears to be determined, in part, by the intrinsic hydrophobic domain of the 32-34 kilodalton polypeptide. The biogenesis of the 32-34 kilodalton polypeptide is discussed with relation to genetic mechanisms that may be responsible for triazine resistance at the level of the chloroplast membrane.

[1] Current address: University of Wurzburg, Botanical Institute, Wurzburg 87, West Germany.

0097-6156/82/0181-0037$05.00/0

Approximately half of all commercial herbicides act by inhibiting photosynthesis by interacting with specific sites along the photosynthetic electron transport chain. A number of diverse chemicals including the ureas, amides, triazines, triazinones, uracils, pyridazinones, quinazolines, thiadiazoles, and certain phenols are thought to act specifically at a common inhibitory site at the reducing side of photosystem II (PS II) (1, 2). Several lines of evidence indicate that this inhibition occurs at the level of a protein bound plastoquinone called "B" (3). This electron carrier acts as the second stable electron acceptor of PS II (4, 5). It has been proposed that the common mode of action of these chemical classes is via high-affinity binding to the PS II complex (6). Herbicide binding induces a change in the redox potential of the quinone cofactor of B, thereby making the transfer of electrons from Q (the first stable electron acceptor of PS II) thermodynamically unfavorable (3, 5).

Establishment of structure-activity relationships within herbicide classes has been extremely useful in elucidating important structural aspects of the inhibitor molecules themselves (7). Until recently, however, much less emphasis has been directed at understanding the biochemical characteristics of the herbicide receptor site within the chloroplast membrane. Progress in this direction was initiated when Strotmann and Tischer (6, 8) introduced techniques for monitoring specific binding of herbicides to isolated chloroplast thylakoid membranes. These and other workers (3, 6, 9) have utilized a variety of radiolabeled inhibitors of PS II function for the characterization of properties of the herbicide binding site; the studies have resulted in the demonstration of a competition for a single binding domain per photosynthetic electron transport chain.

An understanding of the biochemical characteristics of the PS II localized herbicide receptor domain is particularly relevant because of the appearance of triazine-resistant weed biotypes in the United States, Canada, and Europe (3). Initial attempts at understanding the mechanism(s) of resistance directed investigators to evaluate alterations in uptake, translocation, or metabolism of triazines. Only small differences between susceptible and resistant biotypes were established, these being insufficient to explain the mechanism of extreme herbicide resistance.

As the primary mechanism of action of the *s*-triazines involves inhibition of PS II electron transport, attention was also directed at analysis of chloroplast reactions in resistant weed biotypes (10, 11, 12). These studies can be summarized as follows: (a) in all cases studied to date, there is a modification in the chloroplast membranes of resistant biotypes that changes the characteristics of *s*-triazine binding; (b) this modification results in altered binding characteristics of other classes of herbicides, (i.e., only slight resistance to ureas, but increased sensitivity to phenols) (see 13 for review), and (c) the alteration of the herbicide receptor in resistant weeds is accompanied by a detectable change in the kinetic characteristics of electron transfer from Q^- to B in the native membranes even in the absence of all herbicides (3, 14, 15), which indicates that the apoprotein of B may be altered in the resistant chloroplasts so that the bound quinone cofactor exhibits different redox properties.

In this paper, we have summarized our current understanding of the biochemical nature of the triazine binding site within the PS II complex. Studies using the proteolytic enzyme trypsin as a selective, surface-specific modifier of membrane polypeptides and the use of a photoaffinity triazine have been utilized separately to identify the triazine receptor protein as a 32-34 kilodalton (kDal) polypeptide of the PS II complex in peas (*Pisum sativum* L.). The nature of the covalent attachment of the photoaffinity probe has also enabled us to identify the triazine receptor protein as a product of chloroplast-directed protein synthesis; this implies that the structural gene for the triazine receptor polypeptide is encoded on chloroplast DNA. This is in agreement with reports, based on classical genetic analysis, that triazine resistance in *Brassica campestris* L. (16) is a maternally inherited trait.

Materials and Methods

Chloroplast Isolation. Chloroplasts of peas, spinach (*Spinacia oleracea* L.), and biotypes of *Amaranthus hybridus* L. susceptible or resistant to *s*-triazines were isolated and stroma-free thylakoids prepared as previously described (17). Intact chloroplasts were obtained from pea leaves following the method of Blair and Ellis (18).

Trypsin Treatment. Trypsin incubations were carried out at room temperature as previously described (19). Trypsin concentrations used are specified in the text.

Photoaffinity Labeling. The photoaffinity labeling of pea thylakoid membranes using uniformly ^{14}C-labeled 2-azido-4-ethylamino-6-isopropylamino-*s*-triazine (azidoatrazine) was carried out as described (20). The compound was generously provided by Dr. Gary Gardner, Shell Development, Modesto, CA. Synthesis of the compound was as reported previously (21).

Radioactive Amino Acid Incorporation. For analysis of *in vivo* protein synthesis in leaves of pea seedlings and of *A. hybridus*, a surface application of 100-150 µCi ^{35}S-methionine (New England Nuclear, 1200-1400 mCi/µmol) in a 100 µl solution containing 0.1% Tween-80 was performed. Uptake and incorporation of the radiolabel into protein was allowed to proceed for 4 h prior to isolation of the thylakoid membranes.

For analysis of chloroplast-directed protein synthesis, isolated intact chloroplasts (0.5 mg Chl/ml) were incubated with 100 μCi ^{35}S-methionine for 40 min in an illuminated, constant temperature (20C) water bath. Following incubation, chloroplasts were broken in a low osmoticum buffer that contained 20 mM Tricine-NaOH (pH 7.8) and centrifuged at 4,000 x g for 5 min. The resulting pellet, containing stroma-free thylakoids, was washed once in the same buffer prior to trypsin treatment.

Polyacrylamide Gel Electrophoresis and Fluorography. Sodium dodecyl sulfate polyacrylamide gel electrophoresis (SDS-PAGE) was carried out using a 10-17% linear polyacrylamide gradient slab gel with a 1 cm 5% stacking gel that incorporated the buffer system of Laemmli (22). Samples were prepared for electrophoresis and gels were run, fixed, and stained for protein as previously described (23).

Analysis of polyacrylamide gels for azido-^{14}C-atrazine-protein complexes or ^{35}S-methionine incorporation into proteins was accomplished by X-ray fluorography as described (24) using Kodak SB-5 X-ray film.

Photosynthetic Reactions. All photosynthetic reaction assays were conducted with isolated spinach thylakoid membranes in 50 mM sodium phosphate (pH 6.8), 100 mM sorbitol, 10 mM NaCl, 5 mM $MgCl_2$ (PSNM buffer). Gramicidin D (10^{-7} M) and NH_4Cl (1 mM) were included to uncouple electron transport. Photosynthetic inhibitors were prepared as concentrated stock solutions in absolute methanol. Addition volumes to assay media were in all cases less than 0.5% of the final volume.

Measurements of 2,6-dichloroindophenol (DCPIP) photoreduction were carried out as previously described, with a Hitachi Model 100-60 Spectrophotometer modified for cross-illumination (12).

Herbicide Binding Assays. Control and trypsin-treated chloroplast thylakoids were suspended in PSNM buffer. Buffer (1 ml volume) containing 50 μg Chl was incubated 3 min with ^{14}C-atrazine (specific activity: 27.2 μCi/mg). Chloroplasts were pelleted and an aliquot of the supernatant was removed for determination of the amount of unbound atrazine. Details of this procedure are described elsewhere (6, 9). Radiolabeled atrazine was a gift of Dr. H. LeBaron, CIBA-GEIGY, N. Carolina.

Results and Discussion

Modification of Herbicide Binding Sites by the Proteolytic Enzyme Trypsin. The proteolytic enzyme trypsin has been used to selectively modify chloroplast membrane proteins that are exposed at the membrane surface (19, 23). The use of trypsin as a topographic probe has been developed extensively over the last several years to analyze functional components of the PS II complex and to study mechanisms by which inhibitors of PS II function, most notably diuron (DCMU), block electron transport (25, 26, 27).

In summary, it has been established that the reaction center of PS II is relatively stable against trypsin attack, but that one or more components on the reducing side of PS II are accessible to trypsin and are modified by the enzyme in a time-dependent manner, resulting in loss of electron transport activity to such artificial electron acceptors as DCPIP and methyl viologen (19). In addition, trypsin treatment causes herbicide-insensitive electron flow that results from the creation of a new artificial pathway for electron transfer to certain added electron acceptors, such as ferricyanide. This new electron transfer probably arises directly from the primary acceptor Q (19, 26, 27). The results of functional studies using a variety of artificial electron acceptors are summarized schematically in Figure 1. Acceptors that remove electrons from the chain, at or after the level of plastoquinone, are inhibited following trypsin treatment because of the inactivation of an electron carrier acting after Q. Transfer of electrons to the artificial acceptor SiMo is not inhibited initially by trypsin; thus the PS II components that function up to the level of Q are protected from trypsin attack. It appears that the carrier most sensitive to trypsin is the apoprotein of the secondary PS II electron acceptor, B (19).

A direct correlation between trypsin-mediated loss of electron transport through B, and loss of herbicide binding sites can be made by parallel measurement of electron transport and ^{14}C-atrazine binding to trypsin-treated membranes. Such an analysis is summarized in Figure 2A. The number of herbicide binding sites in trypsin-treated thylakoids is compared to the absolute rate of uninhibited noncyclic electron transport ($H_2O \rightarrow$ DCPIP). The time-dependent loss of binding sites for atrazine paralleled the loss of noncyclic electron transport. From this analysis, we conclude that the presence of the herbicide binding protein must be required for function on the reducing side of PS II.

Trypsin-mediated changes in atrazine affinity are compared in Figure 2B by two methods. Trypsin-treated membranes were tested for electron transport in the presence and absence of 0.25 μM atrazine; this concentration of atrazine resulted in 50% inhibition of electron transport activity in control samples. Within the first 5 min of trypsin treatment, a marked decrease in affinity to atrazine was observed as indicated by a decrease in sensitivity of electron transport to atrazine. Measurements of herbbicide binding using radioactive atrazine showed a parallel decrease in affinity of the atrazine for trypsin-treated membranes.

When the time dependence of trypsin-mediated loss of atrazine affinity (Figure 2B) is compared to loss of herbicide binding sites (Figure 2A), two distinct time courses are resolved. The initial response to trypsin attack was a rapid alteration in affinity (within the first 5 min of incubation with the

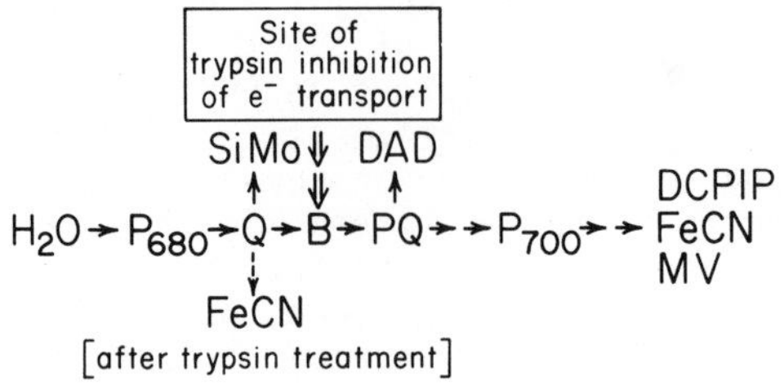

Site of trypsin inhibition of e^- transport

SiMo ⇓ DAD

$H_2O \rightarrow P_{680} \rightarrow Q \rightarrow B \rightarrow PQ \rightarrow\rightarrow P_{700} \rightarrow\rightarrow$ DCPIP / FeCN / MV

FeCN

[after trypsin treatment]

Z. Naturforsch.

Figure 1. Schematic summary of electron acceptor sites in the electron transport chain before and after trypsin treatment.

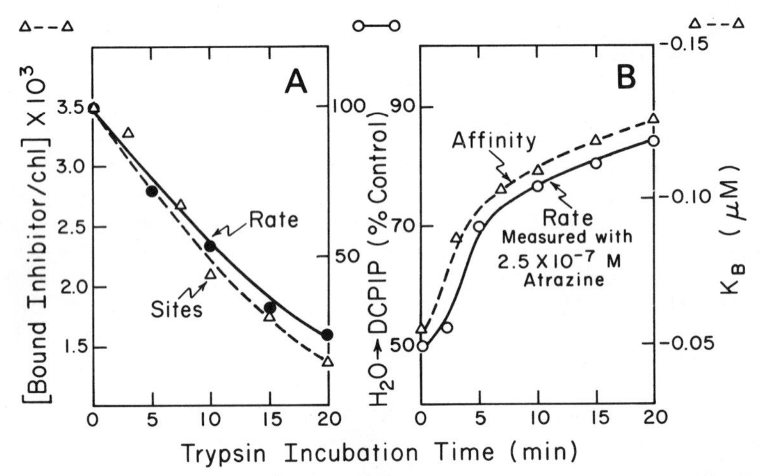

Z. Naturforsch.

Figure 2A. Comparison of data derived from studies of ^{14}C-atrazine binding for thylakoid membranes incubated with 2 μg of trypsin/mL for various times (0–20 min) with that obtained from direct analysis of electron transport capacity ($H_2O \rightarrow$ DCPIP).

Figure 2B. Comparison of data derived from studies of ^{14}C-atrazine affinity for thylakoid membranes incubated with trypsin with that obtained from electron transport assays ($H_2O \rightarrow$ DCPIP) in the presence of 0.25 μM atrazine.

proteolytic enzyme). This was followed by a gradual change in affinity over the next 15 min of treatment. This is in contrast to the gradual time-dependent decrease in binding sites, implying at least a two-step alteration in the protein (or proteins) of the PS II complex that constitute the atrazine binding site.

Mild trypsin treatment has been shown also to alter the affinity for a number of other chemical families of PS II directed herbicides in a manner similar to that of the s-triazines. Trypsin-mediated decreases in inhibitory activity are found for uracil (19), urea, pyridazinone (19, 28) and triazinone (28) herbicides. In contrast, phenol-type herbicides increased in inhibitory activity following brief trypsin treatment (19, 28), although the trend was reversed over longer treatment periods.

The distinctly different behavior of the phenol-type herbicides following trypsin treatment suggests that different determinants within the PS II protein complex establish the "domains" that regulate the binding properties of these inhibitors. In spite of the fact that phenol-type herbicides will displace bound radiolabeled herbicides such as diuron, these inhibitors show noncompetitive inhibition (29, 30). At present, there are three lines of evidence which favor the involvement of two domains within the PS II complex that participate in creating the binding sites for these herbicides: (a) isolated PS II particles can be selectively depleted of a polypeptide with parallel loss of atrazine sensitivity, but not dinoseb inhibition activity (33); (b) in resistant weed biotypes, chloroplast membranes that exhibit extreme triazine resistance have increased sensitivity to the phenol-type herbicides (13); and (c) experiments with azido (photoaffinity) derivatives of phenol and triazine herbicides result in the covalent labeling of different PS II polypeptides (20, 31).

Identification of the Triazine Binding Site. An ultimate goal for the biochemical understanding of herbicidal effects on chloroplast membranes is to identify the specific membrane constituent that serves as herbicide receptor. For the s-triazines, identification of the specific receptor is required for understanding triazine resistance at the molecular level. Because trypsin modification of membranes results in protein-specific alterations of the membrane, the stepwise loss of herbicide binding affinity and binding sites is attributed to the stepwise alteration of a protein that serves as the herbicide receptor. One approach to the identification of herbicide receptors has been the analysis of peptide alterations following trypsin treatment by using SDS-PAGE. Analysis of polypeptide alteration in thylakoid membranes has been unsatisfactory, however, because of the multiplicity of protein changes brought about by trypsin (19). However, characterization of membranes subfractionated following detergent treatment has been somewhat more informative. When PS II enriched subfractions are isolated from trypsin-treated membranes, the number of altered polypeptides that could correspond to an altered receptor protein is narrowed down to four candidates. Among these is a polypeptide of 32 kDal (19). Highly purified PS II complexes isolated by detergent fractionation also contain a polypeptide of 32 kDal (32, 33). Trypsin treatment of isolated PS II complexes results in the degradation of the 32 kDal polypeptide with concomitant loss of herbicide activity (32, 33). Finally, selective removal of the 32 kDal polypeptide from PS II particles by additional detergent treatments results in loss of both diuron and atrazine binding (33). These lines of evidence all implicate a 32 kDal polypeptide as the receptor for triazine and urea classes of herbicides.

The association of diuron and atrazine with chloroplast membranes is via a high-affinity, but noncovalent binding. Attempts at physical isolation of proteins labeled by radioactive inhibitors have failed because techniques such as detergent fractionation or electrophoretic separation rapidly lead to a new equilibrium and the dissociation of the noncovalent receptor-inhibitor complex. One approach that overcomes this difficulty in identifying a herbicide receptor is to attach a radiolabeled photoaffinity azido derivative of the herbicide to its high-affinity receptor protein. Activation of the azido function of photoaffinity compounds by UV irradiation produces a nitrene that is highly reactive (34). If the compound remains localized at its high-affinity site throughout the lifetime of the destabilized nitrene group, covalent binding will occur at the binding site. Azidoatrazine has been shown to inhibit photosynthetic electron transport at a site identical to that of atrazine (20). We have used azidoatrazine as a photoaffinity probe to identify the herbicide-receptor protein in chloroplasts of A. hybridus. In order to demonstrate the specificity of binding to a high-affinity site, membranes from both susceptible and resistant biotypes were utilized. Analysis of membrane polypeptides from susceptible and resistant membranes by SDS-PAGE is shown in Figure 3 (Coomassie blue stained polypeptides are in lanes A). No major differences in polypeptide composition or staining intensity between the two samples are apparent. For both cases, membranes were UV irradiated in the presence of azido-^{14}C-atrazine prior to electrophoresis. Analysis of the gel by a fluorographic technique showed no detectable bound radiolabel associated with the membrane sample from resistant chloroplasts. In the susceptible membranes, however, the radiolabel extended from a region corresponding to 34 kDal to that of a stained polypeptide at 32 kDal. We shall present evidence in the following discussion that this pattern of labeling is due to covalent herbicide association to a single protein which exists in either of two molecular weight species: a developmental precursor form of 34 kDal which is size-processed to a 32 kDal form. For the remaining discussion we shall designate this triazine herbicide receptor as the 32-34 kDal polypeptide.

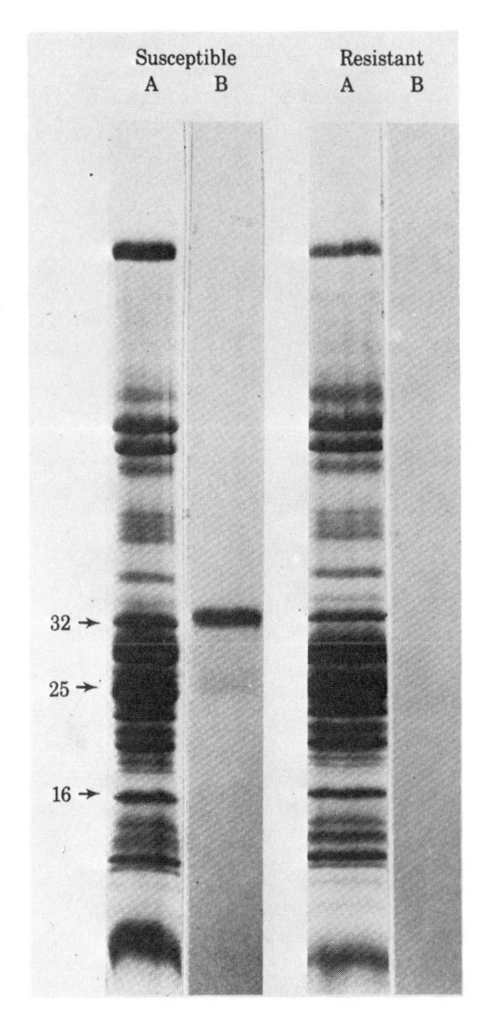

Proc. Natl. Acad. Sci.

Figure 3. Polyacrylamide slab gel electrophoresis of thylakoid membrane polypeptides from susceptible and resistant biotypes of A. hybridus, *stained for protein (lanes A) and by fluorography (lanes B). Susceptible and resistant membranes were incubated with 0.5 μM azido-^{14}C-atrazine under UV light for 10 min prior to SDS solubilization. The predominant location of the radiolabel, as shown by fluorography, is over the 34- to 32-kDal size class.*

One additional line of evidence that independently implicates a 32 kDal polypeptide in triazine binding comes from studies on chloroplasts isolated from a maize mutant that specifically lacks the stainable 32 kDal polypeptide and the rapidly labeled 34 kDal polypeptide. Chloroplasts of this PS II-less lethal mutant lack binding sites for radioactive atrazine (35).

Step-wise Modification of the Herbicide Binding Polypeptide In Vitro. Evidence from atrazine binding studies described in a previous section suggested that trypsin-mediated alterations of surface exposed membrane polypeptides resulted in a sequential alteration of atrazine binding sites. First, a rapid alteration of herbicide affinity was detected followed by a more gradual loss of detectable binding sites. This stepwise alteration was further investigated at the level of polypeptide structure by using the selective membrane modifier, trypsin, against membranes that had been covalently tagged with the radiolabeled photo-affinity triazine probe. As shown in Figure 4A, the use of low concentrations of azido-^{14}C-atrazine resulted in labeling of polypeptide species at 34 kDal. We interpret these data as indicating that the 34 kDal form of the 32-34 kDal polypeptide creates the high-affinity herbicide binding site. In membrane samples treated with 2 µg trypsin/ml for 15 min (see legend of Figure 4 for details), a trypsin concentration shown previously to principally bring about changes in atrazine affinity (Figure 2B), the 34 kDal polypeptide labeled with azidoatrazine was altered to a species that comigrated with a stained 32 kDal polypeptide (Figure 4B). At higher trypsin concentrations (Figures 4C and 4D; 10 and 40 µg trypsin/ml, respectively) where loss of both electron transport function and inhibitor binding sites were observed previously, the polypeptide tagged by azidoatrazine was further degraded to species at 18 and then 16 kDal in a sequential, stepwise manner. [The experiment of Figure 4 utilized membranes which were first labeled with azidoatrazine and then subjected to trypsin treatment.] We conclude that the major covalent attachment site for azidoatrazine is in a hydrophobic region of the membrane which is inaccessible to trypsin, thereby leaving an intrinsic 16 kDal fragment of the triazine binding protein associated with the membrane following trypsin treatment. Further degradation of the 16 kDal polypeptide is not observed after prolonged trypsin treatment.

The data of Figure 4 indicated that a 34 kDal form of the 32-34 kDal polypeptide is the high-affinity binding site for triazines. To test whether the trypsin-derived, membrane-bound fragments of this protein still bound herbicide, azidoatrazine was used against trypsin-treated membranes. Figure 5 shows the fluorogram of electrophoretically separated pea chloroplast polypeptides from control (A and B) and trypsin-treated (C and D) membranes that were tagged with azidoatrazine after the protease modification. When azidoatrazine was applied against

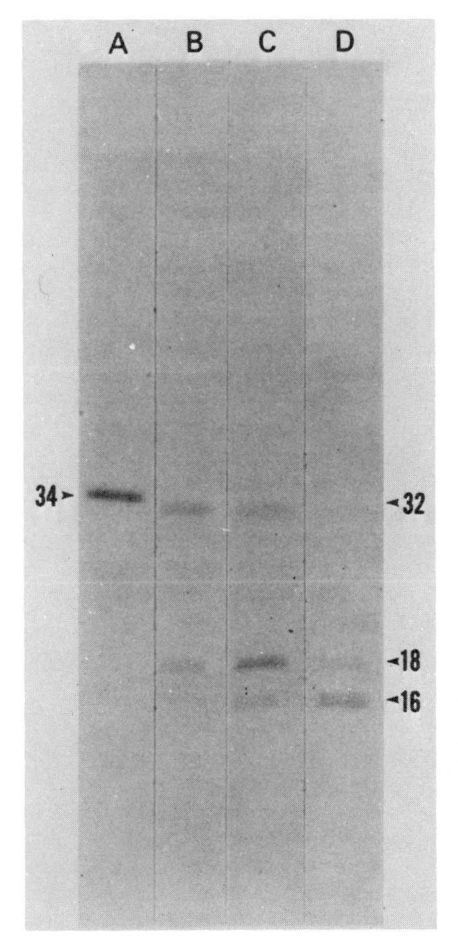

Figure 4. Fluorogram of a polyacrylamide gel that shows trypsin sensitivity of the 34-kDal polypeptide of pea chloroplast membranes following covalent radiolabeling with azido-^{14}C-atrazine. Key: A, control; B, 2 μg trypsin/mL; C, 10 μg trypsin/mL; D, 40 μg trypsin/mL for 15 min at room temperature. Proteolysis was stopped by the addition of a 20-fold excess of trypsin inhibitor. Membranes were washed twice in PSNM buffer prior to electrophoresis.

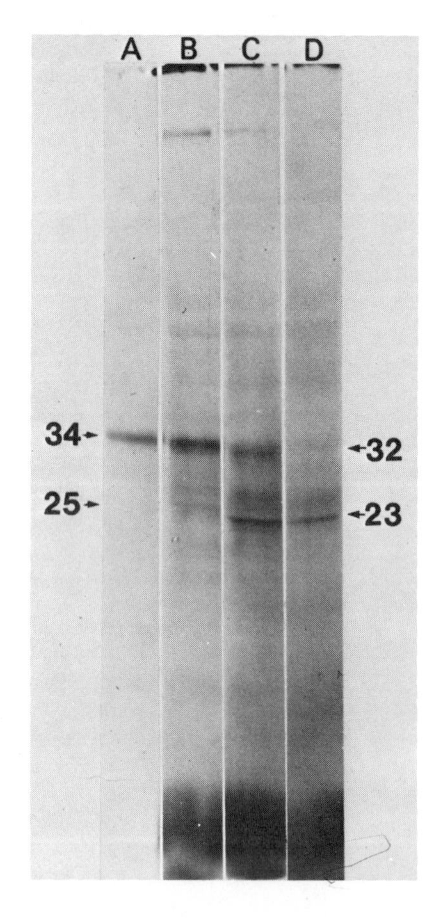

Figure 5. Fluorogram of electrophoretically separated pea chloroplast polypeptides from control and trypsin-treated membranes that were tagged with radiolabeled azidoatrazine. Key: A, control membranes (50 μg Chl/mL) UV-irradiated 10 min in the presence of 2.5 μM azido-^{14}C-atrazine; B, as in A, but with 25 μM azidoatrazine; C, treatment of chloroplasts with 2 μg trypsin/mL for 15 min prior to tagging with 25 μM azidoatrazine; D, as in C, but 40 μg trypsin/mL for 15 min prior to tagging with azidoatrazine.

control membranes at a concentration of 2.5 μM (equivalent to an atrazine concentration required to inhibit electron transport by 90%) only the 34 kDal polypeptide species was labeled (Figure 5A). When a 10-fold higher concentration is utilized (25 μM), equivalent to an atrazine concentration required to inhibit electron transport by 90% in trypsin-treated membranes) both the 34 and 32 kDal forms of the 32-34 kDal polypeptide are labeled (Figure 5B).

Mild trypsin-treatment (2 μg trypsin/ml, 15 min) followed by azidoatrazine tagging using 25 μM azidoatrazine (Figure 5C) resulted in tagging of the 32 kDal species (in contrast to labeling both the 34 and 32 kDal forms of the polypeptide in control membranes). This supports the hypothesis that the presence of the trypsin-derived 32 kDal polypeptide provides both an affinity site and a binding site for triazines. When thylakoids are tagged following high trypsin treatment (40 μg trypsin/ml, 15 min) no label is observed at 32 kDal, nor at the molecular weights of the expected breakdown products of 18 and 16 kDal. These data further suggest that affinity and/or binding to the intrinsic fragments alone can not occur. It should be noted that in all cases where a high concentration of azidoatrazine is utilized (25 μM, Figures 5B, C, and D) a detectable level of nonspecific binding of the compound to a number of thylakoid polypeptides is observed. The most intensely labeled was a polypeptide species of 25 kDal before trypsin treatment or 23 kDal after protease digestion. This polypeptide was previously demonstrated to be the apoprotein of the light-harvesting chlorophyll a/b pigment protein which has a surface-exposed segment (23).

The Triazine Receptor Protein is a Product of Chloroplast-Directed Protein Synthesis. Triazine resistance has been demonstrated from reciprocal crossing experiments to be inherited uniparentally through the female parent in *B. campestris* (16). As the resistance mechanism has been shown to reside at the level of the chloroplast membrane, a role for the chloroplast genome in conferring resistance is strongly implied (17).

It was of interest to determine if the chloroplast membrane protein of 32-34 kDal that binds the photoaffinity triazine, and which appears to be required for triazine binding in isolated PS II particles, is a chloroplast gene product. In developing chloroplasts, in parallel to the appearance of functional activities, there is rapid synthesis and accumulation of a major thylakoid protein of 34 kDal (36). This rapidly synthesized chloroplast protein has been shown to be encoded by the chloroplast genome in *Z. mays* (37). This section outlines experiments that were carried out to determine if the chloroplast polypeptide, which serves as the triazine binding site, is identical to the chloroplast-encoded protein of the same molecular weight.

As was shown in Figure 4, the 34 kDal polypeptide labeled by azidoatrazine shows a specific stepwise alteration mediated by the proteolytic enzyme trypsin. We interpret that the specific stepwise alteration of this polypeptide is regulated by (a) the presence of lysine or arginine residues in the primary sequence of this peptide that are sterically accessible to the proteolytic probe and (b) the increased accessibility of specific cleavage sites caused by trypsin-mediated alterations of other surface-exposed peptides that share a common microenvironment at the membrane surface with the 32-34 kDal polypeptide. The stepwise trypsin degradation of the 34 kDal form of the polypeptide to 32, 18, and 16 kDal defines a "map" of its structural integration within the membrane and can be used to identify the same protein "tagged" by an independent method.

The chloroplast-synthesized protein of the same molecular weight as the triazine-binding protein is the most rapidly synthesized protein *in vivo* of pea thylakoid membranes. Thus, its trypsin sensitivity can be readily evaluated by autoradiographic techniques. Pea seedlings were allowed to take up and incorporate ^{35}S-methionine for 4 h; this resulted in radiolabeling of rapidly synthesized polypeptides. Control and trypsin-treated membrane samples were subjected to SDS-PAGE. Following staining for protein, polypeptides were analyzed for radiolabel incorporation by X-ray fluorography.

As shown in Figure 6A, incorporation of ^{35}S-radiolabel was observed for two major molecular weight species -- the apoprotein of the light harvesting complex (LHC) at 25 kDal and a polypeptide of 34 kDal. Following trypsin treatment (Figure 6B), the radiolabel associated with the LHC polypeptides was altered in electrophoretic mobility by 2 kDal following the expected alteration for the Coomassie-stained peptide species (23). The rapidly synthesized 34 kDal polypeptide was also susceptible to trypsin. Corresponding to its loss with trypsin treatment, new radiolabeled bands appeared at 32 (Figure 6B), then 18, and 16 kDal (Figures 6C,D) in an identical, trypsin-concentration dependent fashion to that of the photoaffinity tagged 34 kDal polypeptide (see Figure 4). From the identical trypsin sensitivity of the photoaffinity tagged polypeptide and the rapidly synthesized chloroplast protein of the same molecular weight, we conclude that the two polypeptides are one and the same.

Evidence that the Triazine Binding Protein is Present in Triazine Resistant Weed Biotypes. The utilization of a photoaffinity labeled triazine herbicide has been invaluable in the definitive identification of one specific polypeptide of the PS II complex as the triazine binding site. The absence of covalent labeling in resistant membranes suggests that either the polypeptide is missing from the membrane or that it is present, but genetically altered, resulting in an alteration in its primary structure, and possibly changing its conformation in the membranes.

In susceptible membranes, triazine and urea herbicides com-

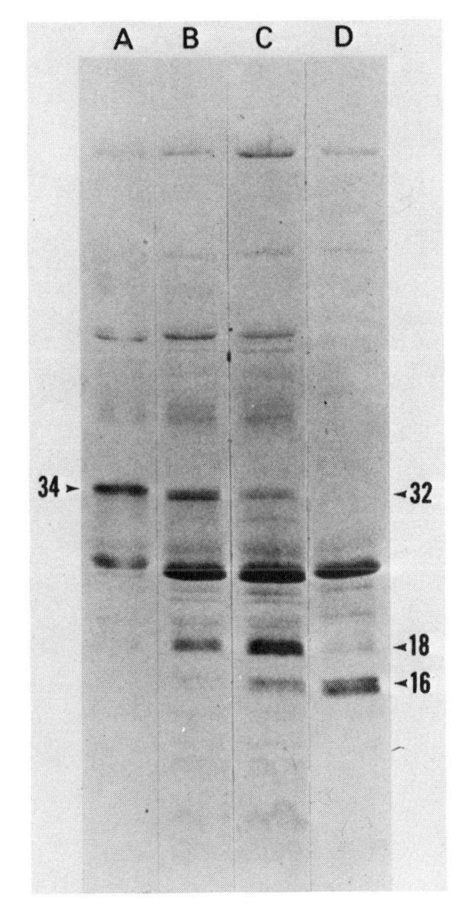

Figure 6. Fluorogram of polyacrylamide gel showing trypsin sensitivity of the 34-kDal polypeptide of pea chloroplast membranes following in vivo incorporation of ^{35}S-methionine in whole leaves. Key: A, control; B, 2 μg trypsin/mL; C, 10 μg trypsin/mL; D, 40 μg trypsin/mL. Treatments were as described in Figure 4.

pete for the same binding region as determined by direct competition studies (9). In resistant membranes, whereas triazine sensitivity is extensively diminished, the ability of diuron to inhibit electron transport is not altered significantly (3, 9, 13). This indicates an alteration in the triazine, but not the urea affinity site and further suggests that a common binding site, i.e., protein, is present in resistant membranes, but possesses an altered affinity for the triazine herbicides. It was of interest to determine if the polypeptide responsible for herbicide sensitivity in chloroplasts of normal, triazine-susceptible plants was also present, i.e., synthesized, in the resistant biotypes. Using the techniques for ^{35}S-methionine incorporation *in vivo* described for peas, incorporation of ^{35}S-radiolabel into thylakoid polypeptides of susceptible and resistant biotypes of *A. hybridus* was investigated. The data shown in Figure 7 demonstrate that the 34 kDal polypeptide is synthesized in both herbicide-susceptible and resistant plants. Furthermore, the polypeptide in resistant membranes shows an identical sensitivity to trypsin-treatment as the 34 kDal polypeptide of the susceptible membranes. Whereas it is evident that the polypeptide is synthesized and present in both biotypes, there is no apparent size difference or change in membrane orientation, as reflected by trypsin sensitivity, that can account for the extreme differences in triazine affinities. The triazine-resistance mechanism, therefore, probably resides in an altered primary structure of this protein.

Acknowledgements: This research was supported, in part, by a Grant from the United States - Israel Binational Agricultural Research and Development Fund (BARD), DOE Contract DE-AC02-7ERO-1338 to Michigan State University, and a Wellesley College Faculty Aid Grant to K. Steinback. We also thank Dr. Gary Gardner of Shell Agricultural Chemicals for his collaboration and consultations in experiments using azido-atrazine.

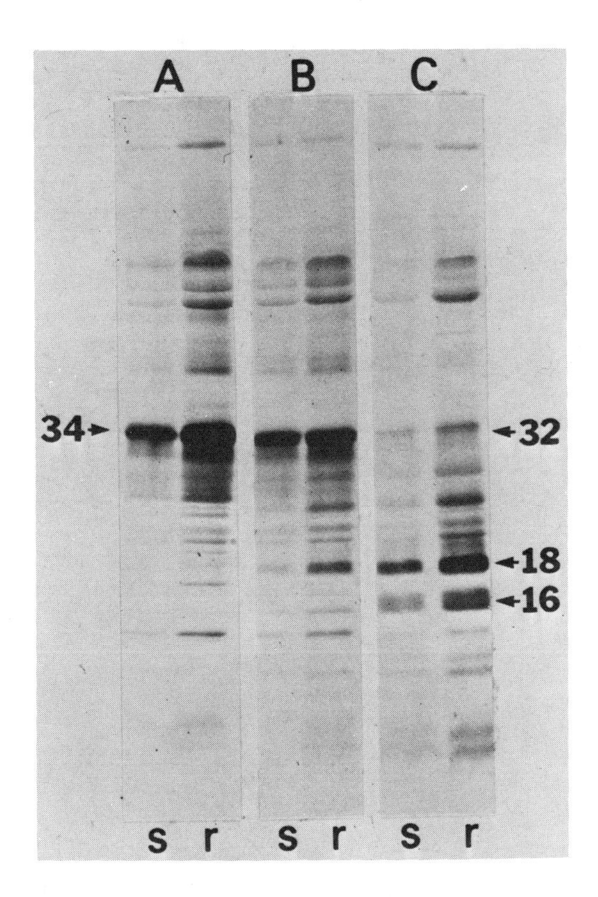

Figure 7. Fluorogram of a polyacrylamide gel showing trypsin sensitivity of the 34-kDal polypeptide of chloroplast thylakoid membranes isolated from susceptible (S) and resistant (R) biotypes of A. hybridus *following in vivo incorporation of ^{35}S-methionine in whole leaves. Isolated thylakoid membranes were treated with A, no trypsin; B, 2 μg trypsin/mL; C, 20 μg trypsin/mL for 15 min as described for Figure 4.*

Literature Cited

1. Ashton, F. M.; Crafts, A. S. "Mode of Action of Herbicides"; Wiley: New York, 1973; pp. 69-99.
2. Wright, K.; Corbett, J. K. Z. Naturforsch. 1979, 34c, 966-72.
3. Pfister, K.; Arntzen, C. J. Z. Naturforsch. 1979, 34c, 996-1009.
4. Bouges-Bocquet, B. Biochim. Biophys. Acta 1973, 314, 250-6.
5. Velthuys, B. R.; Amesz, J. Biochim. Biophys. Acta 1974, 333, 85-94.
6. Tischer, W.; Strotmann, H. Biochim. Biophys. Acta 1977, 460, 113-25.
7. Trebst, A.; Draber, W. "Advances in Pesticide Research"; Geissbuhler, H., Ed.; Pergamon Press: Oxford, 1979; pp. 223-34.
8. Strotmann, H.; Tischer, W.; Edelman, K. Ber. Deutsch Bot. Ges. 1974, 87, 457-63.
9. Pfister, K.; Radosevich, S. R.; Arntzen, C. J. Plant Physiol. 1979, 64, 995-9.
10. Radosevich, S. R.; DeVilliers, O. T. Weed Sci., 1976, 24, 229-32.
11. Souza-Machado, V.; Arntzen, C. J.; Bandeen, J. D.; Stephenson, G. R. Weed Sci., 1978; 26, 318-22.
12. Arntzen, C. J.; Ditto, C. L.; Brewer, P. Proc. Natl. Acad. Sci., U.S.A., 1979, 76, 278-82.
13. Pfister, K.; Watson, J.; Arntzen, C. J. Weed Sci. 1981 (in press).
14. Bowes, J.; Crofts, A. R.; Arntzen, C. J. Arch. Biochem. Biophys. 1980, 200, 303-8.
15. Arntzen, C. J.; Pfister, K.; Steinback, K. E. "Studies of Herbicide Resistance", LeBaron, H.; Gressel, J., Eds.; CRC Press (in press).
16. Souza-Machado, V.; Bandeen, J. D.; Stephenson, G. R.; Lavigne, P. Can J. Plant Sci. 1978, 58, 977-81.
17. Darr, S.; Souza Machado, V.; Arntzen, C. J. Biochim. Biophys. Acta 1981, 634, 219-28.
18. Blair, G. E.; Ellis, R. J. Biochim. Biophys. Acta 1973, 319, 223-34.
19. Steinback, K.E.; Pfister, K.; Arntzen, C. J. Z. Naturforsch. 1981, 36c, 98-108.
20. Pfister, K.; Steinback, K. E.; Gardner, G.; Arntzen, C. J. Proc. Natl. Acad. Sci., U.S.A. 1981, 78, 981-5.
21. Gardner, G. Science 1981, 211, 937-40.
22. Laemmli, U. K. Nature (London) 1970, 227, 680-5.
23. Steinback, K. E.; Burke, J. J.; Arntzen, C. J. Arch. Biochem. Biophys. 1979, 195, 546-57.
24. Lasky, R. A.; Mills, A. D. Eur. J. Biochem. 1975, 56, 335-41.
25. Regitz, G.; Ohad, I. "Proc. 3rd. Int. Congr. on Photosynthesis", Vol. III; Avron, M., Ed.; Elsevier: Amsterdam, 1975; p. 1615.

26. Renger, G. Z. Naturforsch. 1979, 34c, 1010-14.
27. Trebst, A. Z. Naturforsch. 1979, 34c, 986-91.
28. Böger, P.; Kunert, K. J. Z. Naturforsch. 1979, 34c, 1015-25.
29. Reimer, S.; Link, K.; Trebst, A. Z. Naturforsch. 1979, 34c, 419-26.
30. Oettmeier, W.; Masson, K. Pestic. Biochem. Physiol. 1980, 14, 86-97.
31. Oettmeier, W.; Masson, K.; Johanningmeier, U. FEBS Lett. 1980, 118, 267-70.
32. Croze, E.; Kelly, M.; Horton, P. FEBS Lett. 1979, 103, 22-6.
33. Mullet, J. E.; Arntzen, C. J. Biochim. Biophys. Acta 1981, 635, 236-48.
34. Bayley, H.; Knowles, J. R. "Methods in Enzymology", Vol. XLVI; Jackoby, W. B.; Wilchek, M., Eds.; Academic Press: New York, 1977; 69-114.
35. Leto, K.; Keresztes, A.; Arntzen, C. J. Plant Physiol. 1981, (in press).
36. Grebanier, A. E.; Steinback, K. E.; Bogorad, L. Plant Physiol. 1979, 63, 436-9.
37. Bedbrook, J. R.; Link, G.; Coen, D. M.; Bogorad, L.; Rich, A. Proc. Natl. Acad. Sci., U.S.A. 1978, 75, 3060-4.

RECEIVED August 14, 1981.

4

The Role of Light and Oxygen in the Action of Photosynthetic Inhibitor Herbicides

ALAN D. DODGE

University of Bath, School of Biological Sciences, Claverton Down, Bath, Avon BA2 7AY, United Kingdom

Phytotoxic symptoms of injury caused by herbicides that inhibit chloroplast electron transport, such as the ureas, triazines, and hydroxybenzonitriles, are promoted by both light and oxygen. When carbon dioxide fixation is prevented, excess excitation energy leads to the generation of longer lived triplet chlorophyll. If unquenched by carotenoids, this may directly induce proton abstraction from unsaturated fatty acids or singlet oxygen might be generated which induces lipid peroxide formation. Lipid peroxidation leads to cellular destruction and death. Herbicides that divert photosynthetic electron transport, such as the bipyridyls paraquat and diquat, yield the superoxide anion. Superoxide levels produced overtax the normal defense mechanisms. More toxic species such as hydroxyl free radicals are also probably produced which instigate lipid peroxidation and lead to cellular disorganization and death.

James Franck wrote in 1949 (1) "... it is one of the miracles of photosynthesis that the plant can use a dye able to fluoresce in the presence of oxygen, predominantly for the purpose of reduction, and is able to hold the process of photooxidation in check so that damage is prevented or minimized even under severe conditions". It is evident that the chloroplast is endowed with protective devices that are able to limit damage except under extreme conditions. Such situations are promoted by the presence of photosynthetic inhibitor herbicides.

In the normal chloroplast, light energy (hv) absorbed by pigments and in particular by the chlorophylls (Chl) causes excitation to the singlet state. This short-lived state ($\sim 10^{-8}$ sec) is quenched by rapid energy transfer to chlorophyll in the reaction centers. If unquenched, intersystem crossing may lead to the generation of the longer lived triplet state ($\sim 10^{-3}$ sec).

0097-6156/82/0181-0057$05.00/0

$$Chl + h\nu \longrightarrow {}^1Chl \xrightarrow{i.s.c.} {}^3Chl$$

Triplet chlorophyll can be harmlessly quenched by chloroplast carotenoid pigments (2) dissipating the excitation energy (3). If triplet chlorophyll is unquenched, energy transfer from this pigment to ground state oxygen may generate singlet oxygen ($^1O_2$). This is generally termed a type 2 reaction.

$$^3Chl + {}^3O_2 \longrightarrow Chl + {}^1O_2$$

Singlet oxygen may also be quenched by cartenoids (4) and by free radical scavengers located within the chloroplast thylakoids, such as α-tocopherol (5). During active photosynthesis, oxygen concentrations within the chloroplast may be higher than the surrounding cytoplasm (6) and electron leakage from the thylakoids could yield the superoxide anion radical ($O_2^{\cdot -}$) (7). It is possible that this could be formed during normal pseudocyclic electron transport (8). Although the superoxide radical is apparently less toxic than singlet oxygen, its toxicity may be related to the generation of more toxic species such as hydroxyl free ($OH^{\cdot}$) radicals. The chloroplast possesses an efficient superoxide scavenging system in the form of Cu-Zn superoxide dismutase (SOD) (9) to protect against this potential toxic species.

$$O_2^{\cdot -} + O_2^{\cdot -} + 2H^+ \xrightarrow{SOD} H_2O_2 + O_2$$

The dismutating activity of bound manganese (10) may also be important. The generation of hydrogen peroxide in this reaction may be tolerated by the presence of additional enzyme systems that catalyse the destruction of H_2O_2 (11): (i) ascorbate peroxidase, (ii) dehydroascorbate reductase, and (iii) glutathione reductase.

H_2O_2 → H_2O / Ascorbate → Dehydroascorbate (i)

Dehydroascorbate → Ascorbate / glutathione (red) → glutathione (OX) (ii)

glutathione (OX) → glutathione (red) / NADPH → $NADP^+$ (iii)

The potential hazards of oxygen and light might seem to be exacerbated by low carbon dioxide concentrations within the chloroplast. This might be minimized by photorespiration (12) whereby under low carbon dioxide concentrations, oxygen reacts with the carbon dioxide acceptor molecule ribulose bisphosphate to form 3-phosphoglycerate and 2-phosphoglycollate. Further metabolism of phosphoglycollate recycles carbon dioxide to the chloroplast (13).

Electron Transport Inhibitors

A large number of herbicides including the phenylureas, *s*-triazines, uracils, triazinones, hydroxybenzonitriles, pyridazinones, and acylanilides inhibit photosynthetic electron transport (Hill Reaction). The site of action is generally thought to be a protein component located on the outside of the thylakoid membrane (14) and affecting a point between the hypothetical Q and B components of the electron transport chain. As a result of this interaction, electron transport ceases. This leads to a rapid inhibition of carbon dioxide fixation that proceeds at a similar rate irrespective of whether the leaves are incubated in darkness or light (15) (Table I). An interruption of photosynthesis will eventually lead to a reduction of food reserves and subsequent starvation. The appearance of phytotoxic symptoms, such as chlorophyll bleaching, is clearly promoted by light (15, 16, 17) (Figure 1). The inhibition of electron flow will prevent photosynthetic oxygen evolution, but the chloroplast envelope probably provides little resistance to inward oxygen diffusion. Experiments have demonstrated that with both monuron and ioxynil, incubation of cotyledons under argon reduced, but did not prevent the rapid destruction of chlorophyll (Figure 2) (18).

Promotion of the toxic action of photosynthetic inhibitor herbicides by light may be initially ascribed to type 1 reactions. Triplet chlorophyll, without the involvement of oxygen, may directly initiate electron or hydrogen abstraction from particularly susceptible molecules (e.g., unsaturated fatty acids (LH)) to yield lipid free radicals.

$$LH \longrightarrow L^{\cdot}$$

Once initiated, subsequent propagation reactions involving oxygen may generate more free radicals that attack unsaturated lipids in a chain reaction of gathering momentum.

$$L^{\cdot} + O_2 \longrightarrow LO_2^{\cdot}$$

$$LO_2^{\cdot} + LH \longrightarrow L^{\cdot} + LOOH$$

The production of lipid hydroperoxides (LOOH) also may be achieved by the direct interaction of singlet oxygen with unsaturated fatty acids (19).

$$LH + {}^1O_2 \longrightarrow LOOH$$

TABLE I

Carbon dioxide fixation by flax cotyledons measured by an IRGA with illumination of 115 Wm^{-2}, previously incubated with $10^{-3}M$ monuron in the dark or light of 5.25 or 30 Wm^{-2}, from Pallett and Dodge (15)

Incubation Time (Min)	μmol CO_2/g.F.Wt./h		
	Dark	5.25 Wm^{-2}	30 Wm^{-2}
0	68.0	68.0	68.0
15	18.6	11.1	15.4
30	3.2	3.5	2.2
60	2.3	1.9	0.9
120	0.0	0.0	0.0

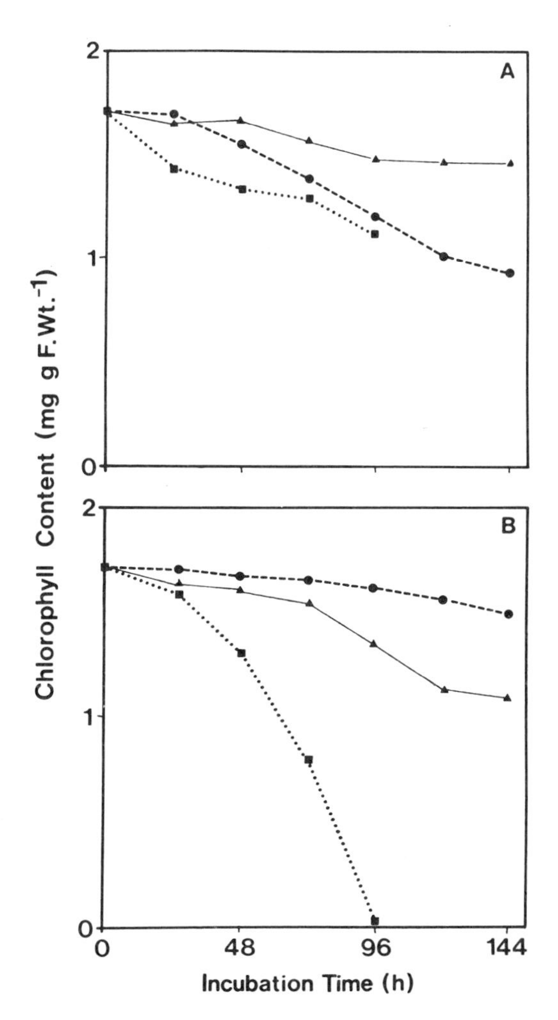

Figure 1. The chlorophyll content of flax cotyledons incubated on water (A) or 10^{-3} M *monuron (B), in the dark (●) or with light of 5.25 Wm^{-2} (▲) or 30 Wm^{-2} (■) (15).*

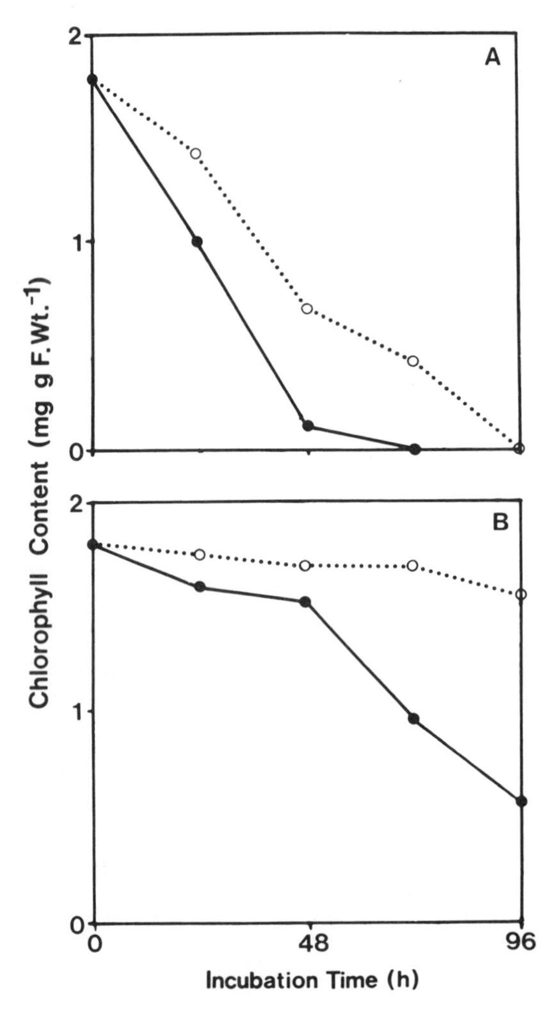

Figure 2. The effect of a 10^{-3} M solution of ioxynil (A) or monuron (B) on the chlorophyll content of flax cotyledons incubated in sealed conical flasks containing air (●) or argon (○) with light of 30 Wm^{-2} (18).

Chloroplast thylakoid membranes are composed of an almost equal percentage of protein and lipid. The acyl lipids of the chloroplast membrane are highly unsaturated; for example, around 80% of the fatty acid component is the 18:3 unsaturated linolenic acid. The consequences of lipid peroxidation reactions are (a) the promotion of membrane destruction which in part may be visibly demonstrated as chlorophyll loss, and observed by electron microscopy as a disorganization of thylakoids and other cellular membranes (18); and (b) the lipid hydroperoxides undergo fragmentation producing short chain hydrocarbons such as ethane (20, 21). Experiments with both monuron and ioxynil showed that the loss of chlorophyll in flax cotyledons was preceded by the breakdown of carotenoid pigments, especially the carotenes, suggesting that this protective system was over-taxed and destroyed. This was followed by a rapid formation of ethane (18) (Figures 3 and 4). Experiments in which monuron treated leaves were treated with the singlet oxygen quencher DABCO (diazobicyclo-octane) showed a limited control of chlorophyll breakdown (22) (Table II).

Many experiments into the secondary effects of photosynthetic inhibitor herbicides have been performed with isolated chloroplasts. Isolated chloroplasts provide a convenient system for studying the generation and quenching of singlet oxygen. Recent experiments with pea chloroplasts illuminated in the absence of an electron acceptor have shown that both chlorophyll and linolenic acid breakdown was retarded by the singlet quenchers DABCO and crocin (23). Further work showed that linolenic acid breakdown and ethane generation in isolated chloroplast thylakoids was promoted by the addition of the singlet oxygen generator rose bengal immobilized on DEAE-sepharose (24).

Divertors of Electron Transport

The interaction of bipyridyl herbicides (paraquat and diquat) with photosynthesis is different from that of the electron transport inhibitors. These compounds, with highly negative redox potentials (paraquat E_0' - 446mV; diquat E_0' - 349mV), interact in the vicinity of ferredoxin causing a diversion of electron flow from the ultimate electron acceptor NADP . This was clearly seen in paraquat-treated plant material as a progressive inhibition of carbon dioxide uptake (25) (Figure 5). Although carbon dioxide uptake rapidly ceased, electron flow from water continued for some time until this system was totally inactivated by the general destruction of cellular integrity (25). Paraquat (and/or diquat) is reduced by a one electron transfer to give the paraquat free-radical. Under anaerobic conditions this radical can accumulate (26, 27). However, in the presence of oxygen, which should be plentiful in the vicinity of the thylakoid in the early stages of paraquat

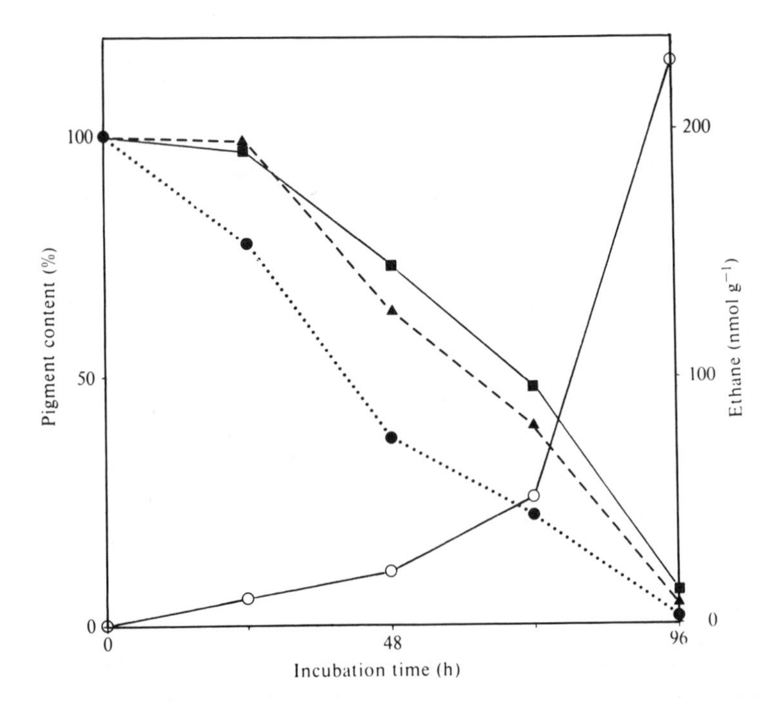

Figure 3. The effect of 10^{-3} M *monuron on pigment content and ethane generation of flax cotyledons incubated at 30 Wm^{-2}. Pigment levels are expressed as a percentage of the levels at 0 h. Key: carotenes, ●; xanthophylls, ▲; chlorophyll, ■; and ethane generation, ○ (18).*

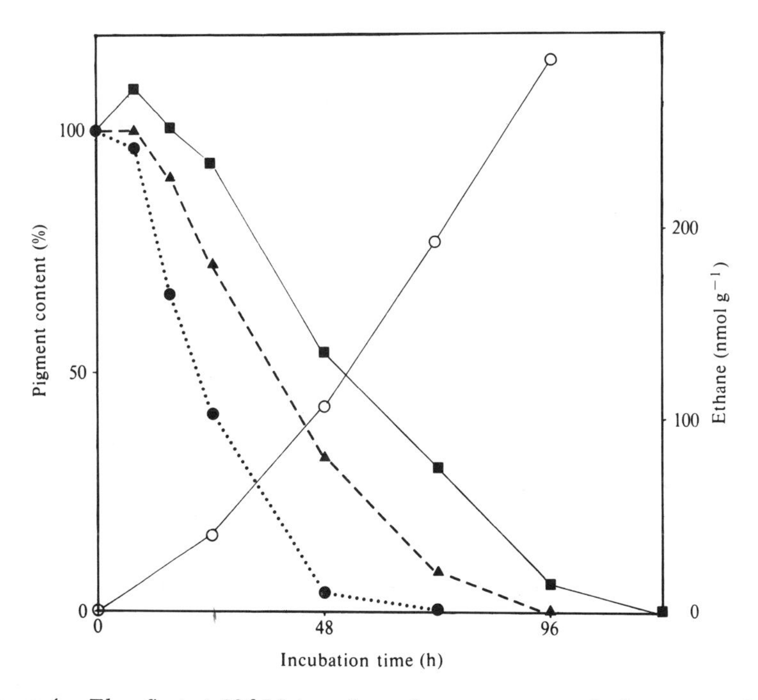

Figure 4. The effect of 10^{-3} M ioxynil on pigment content and ethane generation of flax cotyledons incubated at 5.25 Wm^{-2}. Pigment levels are expressed as a percentage of the levels at 0 h. Key: carotenes, ●; xanthophylls, ▲; chlorophyll, ■; and ethane generation, ○ (18).

TABLE II

The effect of DABCO (10^{-3}M) on the chlorophyll content of flax cotyledons incubated on monuron (10^{-3}M) for 96 h at 5.25 Wm^{-2}, from Youngman *et al*. (22)

Treatment	mg chlorophyll/g. fresh wt.
Water	1.735
Monuron	1.184
DABCO	1.610
Monuron + DABCO	1.598

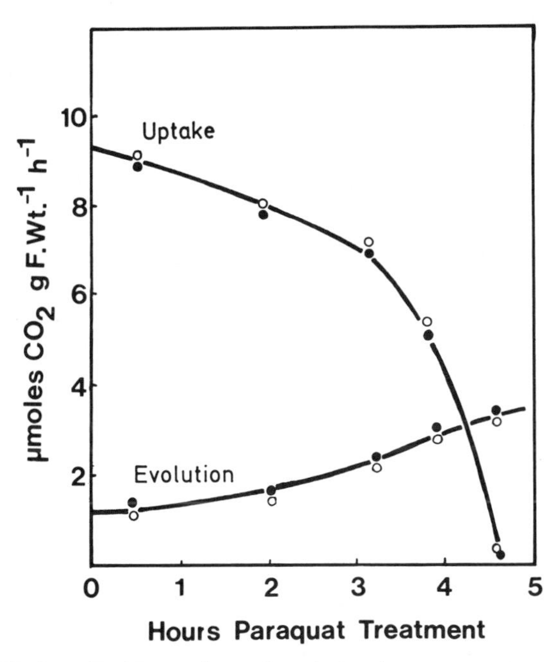

Figure 5. Carbon dioxide uptake and evolution by paraquat-treated flax cotyledons, incubated on paraquat (10^{-4} M) in either light of 5.25 Wm^{-2} (○) or darkness (●). The uptake and evolution of the cotyledons were estimated after treatment (25).

treatment, the reduced paraquat was rapidly reoxidized to yield superoxide :

$$PQ^{2+} + e^{-} \longrightarrow PQ^{\cdot +}$$

$$PQ^{\cdot +} + O_2 \longrightarrow PQ^{2+} + O_2^{\cdot -}$$

The activity of these herbicides is known to be promoted by oxygen (28, 29, 30). It has also been demonstrated that chlorosis is promoted by increasing light (see Figure 11 for *Conyza*) but diminished in the presence of an electron transport inhibitor (28) (Figure 6).

Additional lines of evidence further indicate the importance of superoxide in the action of these herbicides. It has been shown with isolated chloroplasts that the production of superoxide, as measured by the conversion of hydroxylamine to nitrite, was promoted by paraquat and restrained by superoxide dismutase (21) (Figure 7). The generation of superoxide was limited if an electron transport inhibitor such as monuron was present (Table III). Although the chloroplast contains superoxide dismutase enzymes, it is assumed that the rapid generation of $O_2^{\cdot -}$ overtaxes the capabilities of the enzymes. Further experiments with whole leaves and the superoxide scavenger copper penicillamine (31) showed that the presence of this compound not only mitigated the action of the herbicide in promoting chlorophyll decay (32) (Figure 8), but also limited lipid peroxidation as shown by the diminished release of ethane (33) (Figure 9). The feasibility of the superoxide radical being the initial toxic species in vivo was predicted by Farrington et al., (34) using pulse radiolysis studies. It was postulated that the concentration of superoxide within the plant cell could remain constant at 1 μM up to 10 μm or more from the chloroplast thylakoid membrane.

Although it was thought initially that superoxide itself could interact with membrane lipids to initiate lipid peroxidation, more recent evidence has expressed doubt on the potential reactivity of this molecule to induce this process (35). Much discussion has centered on the possible involvement of Fenton and Haber-Weiss type reactions occurring in vivo and generating more toxic species such as hydroxyl free radicals and singlet oxygen. At present it is not possible to affirm that such systems operate (36, 37). However, it might be possible for the reduced paraquat radical to interact with superoxide to generate hydrogen peroxide, a well known product of reduced paraquat reoxidation (38).

$$PQ^{\cdot +} + O_2^{\cdot -} \longrightarrow PQ^{2+} + O_2^{--} \xrightarrow{2H^+} H_2O_2$$

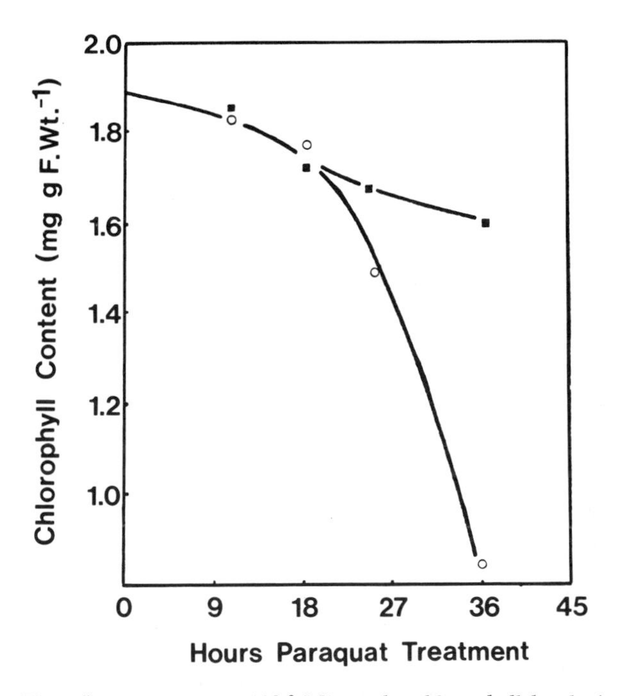

Figure 6. The effect of monuron (10^{-3} M) on the chlorophyll level of paraquat-treated (10^{-4} M) cotyledons under light of 5.25 Wm^{-2}. Key: paraquat, ○; and paraquat plus monuron, ■.

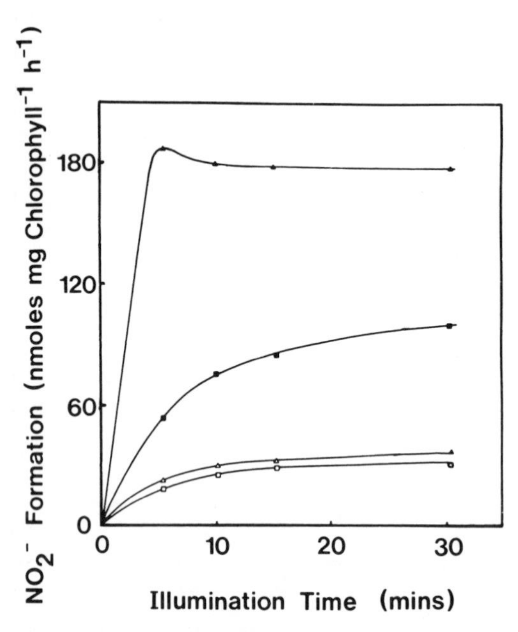

Figure 7. Nitrite formation from hydroxylamine. Reaction mixture contained in 3 mL: 50 mM phosphate buffer pH 7.8; 1 μmol of NH_2OH; chloroplasts with 75 μg of chlorophyll and where indicated 6.6 μM paraquat; 50 units SOD. Illuminated at 275 Wm^{-2} at 22°C. Key: control, ■; plus paraquat, ▲; plus SOD, □; and plus paraquat and SOD, △ (22).

TABLE III

The effect of monuron (10^{-4}M) on the ability of paraquat to promote the formation of nitrite from hydroxylamine in isolated chloroplasts. For further details see legend to Figure 7, from Youngman et al. (32)

Treatment	Water	Monuron
	(nmoles nitrite/mg chlorophyll/hr)	
Control	100	17
Paraquat	195	28

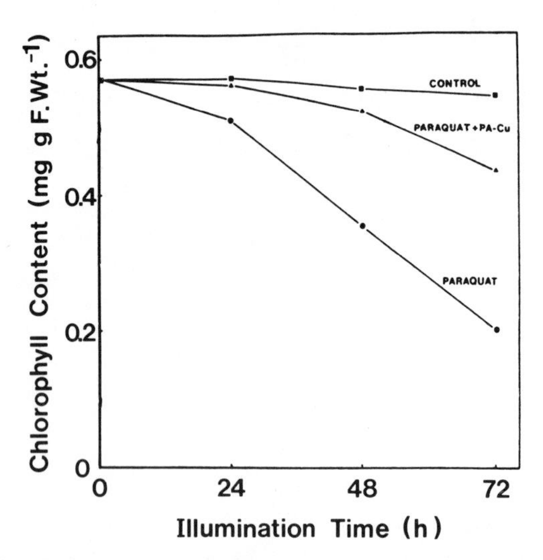

Figure 8. The chlorophyll content of paraquat-treated flax cotyledons incubated in the presence or absence of copper–penicillamine (PA–Cu) under light of 5.25 Wm^{-2}. The final concentration of paraquat was 10^{-5} M *and the PA–Cu represented 50 units of superoxide dismutase. Key: control,* ■; *paraquat plus PA–Cu,* ▲; *and paraquat,* ● *(32).*

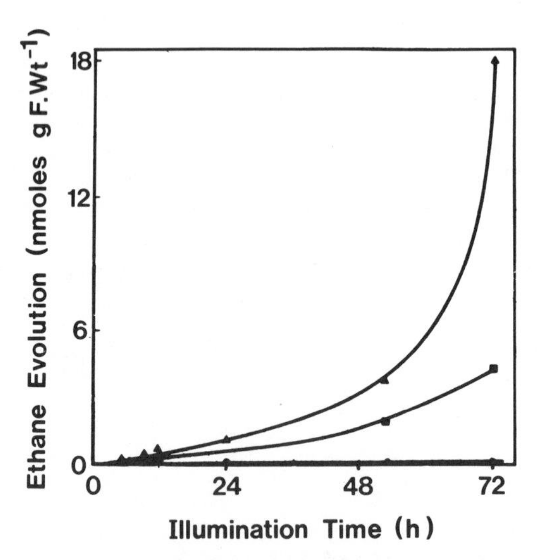

Figure 9. Ethane generation by flax cotyledons incubated in flasks under light of 5.25 Wm^{-2}. The final concentration of paraquat was 10^{-6} M *and the copper–penicillamine (PA–Cu) represented 50 units of superoxide dismutase. Key: control,* ●; *paraquat,* ▲; *and paraquat plus PA–Cu,* ■ *(33).*

Further interaction of reduced paraquat with hydrogen peroxide could yield the more reactive hydroxyl free radical (39) which could initiate lipid peroxidation in membrane fatty acids.

$$PQ^{\cdot+} + H_2O_2 \longrightarrow PQ^{2+} + OH^- + OH^\cdot$$

Electron microscope studies of paraquat-treated leaves showed that after a few hours, membranes had been disrupted and cellular compartmentalization was destroyed. Experiments demonstrating the release of potassium from treated tissue as an indication of plasmalemma and tonoplast disruption (25) (Figure 10) corresponded in time with the visible demonstration of cellular disorganization (40). If the initial action of the herbicides is to indirectly promote lipid peroxidation of membrane unsaturated fatty acids and lead to cellular disorganization, then subsequent deteriorative changes will occur because of the release of vacuolar contents. Not only will there be a rapid change in the osmotic properties of the cell, but the release of vacuolar hydrolytic enzymes will cause further damage. In addition, type 1 and type 2 sensitized reactions may also increase and collectively these changes will lead to rapid plant death.

Although paraquat is used as a broad spectrum total-kill herbicide, tolerance has been found in certain lines of *Lolium perenne* (41) and the weed *Conyza* (42) (Figure 11). In both instances, tolerant lines showed a lack of effect on carbon dioxide fixation, indicating reduced penetration of herbicide (42, 43) (Figure 12). Tolerant lines of *Lolium perenne* were nevertheless shown to possess greater superoxide dismutase activity as well as greater catalase and peroxidase activities (44). Tolerant *Conyza* biotypes showed approximately three-fold increases in superoxide dismutase activity (42).

Conclusion

The phytotoxic action of both electron transport inhibitor and electron deviator herbicides is promoted by light and is largely a response to the overtaxing of protective systems. In both instances, type 1 reactions, implicating the direct action of excited triplet chlorophyll, may be involved. However, of greater potential importance is the interaction of active oxygen species. With electron transport inhibitors, singlet oxygen is produced as an incidental response to unquenched triplet chlorophyll. With electron deviator herbicides, the generation of superoxide is a direct consequence of the diversion of electron flow (Figure 13). Once membrane disruption is

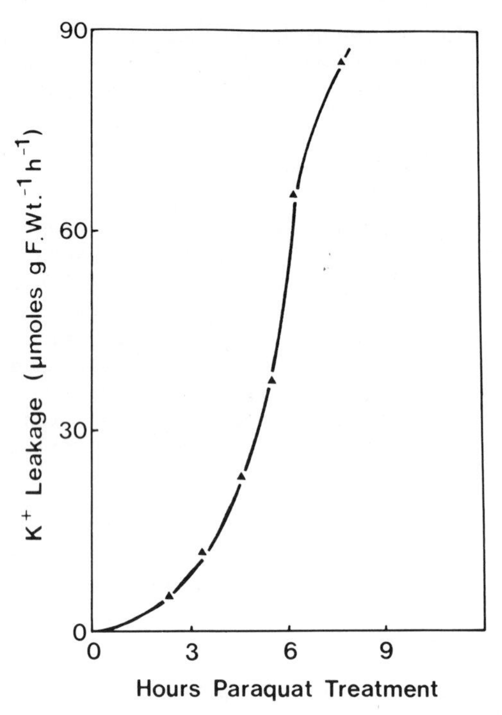

Figure 10. Membrane permeability measured as potassium leakage from slices of paraquat-treated (10^{-4} M*) flax cotyledons (*25*).*

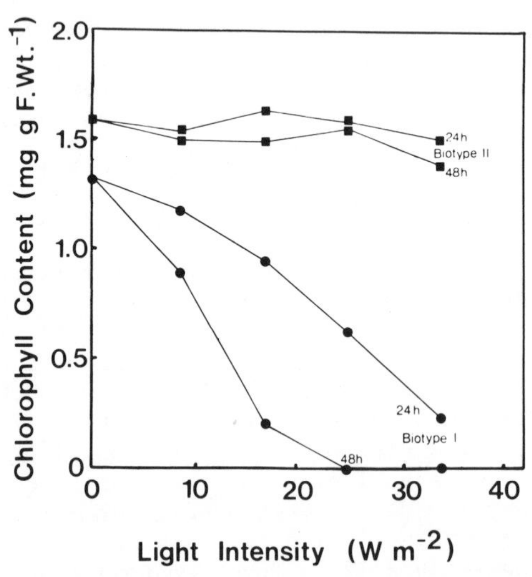

Figure 11. The chlorophyll content of leaf sections of two biotypes of Conyza *treated for 24 or 48 h with paraquat at 10^{-5}* M*, at various light intensities (*41*).*

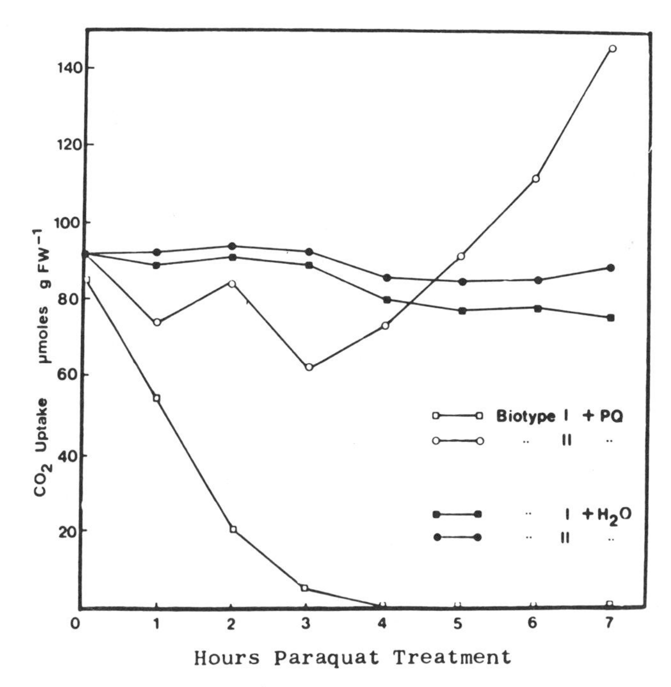

Figure 12. Carbon dioxide exchange of leaf sections of two biotypes of Conyza *treated with or without paraquat of 10^{-5}* M *at 5 Wm^{-2} (41).*

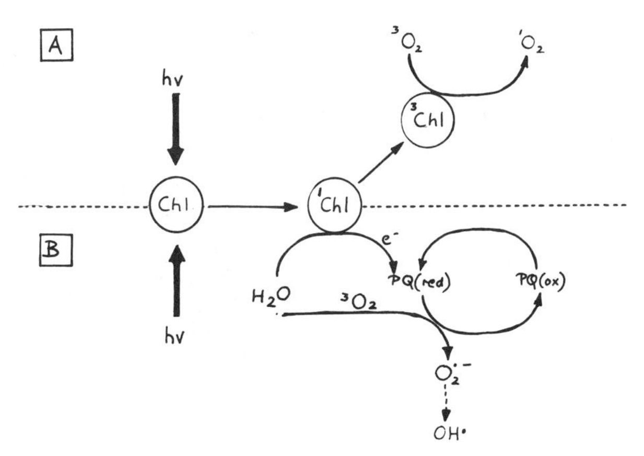

Figure 13. Summary scheme of the primary effects caused by A, photosynthetic inhibitor herbicides and B, photosynthetic deviator herbicides. $^3O_2$ is triplet or ground state oxygen; PQ represents paraquat.

initiated by lipid peroxidation, rapid cellular disorganization and death follows.

Literature Cited

1. Franck, J. "Photosynthesis in Plants"; The Iowa State College Press: Iowa, 1949; p 293.
2. Krinsky, N.I. Pure and Appl. Chem. 1979, 51, 649.
3. Wolff, Ch.; Witt, H.T. Z. Naturforsch. 1969, 246, 1031.
4. Foote, C.S.; Denny, R.W. J. Am. Chem. Soc. 1968, 90, 6233.
5. Hughes, C.T.; Gaunt, J.K.; Laidman, D.L. Biochem. J. 1971, 124, 9p.
6. Steiger, H.M.; Beck, E.; Beck, R. Plant Physiol. 1977, 60, 903.
7. Marsho, T.V.; Behrens, P.W.; Radmer, R.J. Plant Physiol. 1979, 64, 656.
8. Elstner, E.F.; Stoffer, C.; Heupel, A. Z. Naturforsch. 1975, 30C, 53.
9. Asada, K.; Urano, M.; Takahashi, M. Eur. J. Biochem. 1973, 36, 257.
10. Foyer, C.H.; Hall, D.O. FEBS Lett. 1979, 101, 324.
11. Halliwell, B.; Foyer, C.H.; Charles, S.A. Proc. 5th Int. Congr. Photosynth.: Halkidiki, Greece, 1981; in press.
12. Heber, V.; Krause, G.H. TIBS 1980, 5, 32.
13. Chollet, R. TIBS, 1977, 2, 155.
14. Renger, G. Biochim. Biophys. Acta 1976, 440, 287.
15. Pallett, K.E.; Dodge, A.D. Z. Naturforsch. 1979, 34C, 1058.
16. Minshall, W.H. Weeds, 1957, 5, 29.
17. Oorschot, J.L.P. Van; Leeuwen, P.H. Van. Weed Res. 1974, 14, 81.
18. Pallett, K.E.; Dodge, A.D. J. Exp. Bot. 1980, 31, 1051.
19. Rawls, H.R.; Santen, P.J. Van. J. Am. Oil. Chem. Soc. 1970, 47, 121.
20. Riely, C.A.; Cohen, G.; Liebermann, M. Science 1974, 183, 208.
21. Elstner, E.F.; Konze, J.R. Nature 1976, 263, 351.
22. Youngman, R.J.; Pallett, K.E.; Dodge, A.D. "Chemical and biochemical aspects of superoxide and superoxide dismutase". Elsevier/North Holland: New York, 1980; p 402.
23. Percival, M.; Dodge, A.D. (in preparation).
24. Percival, M.; Dodge, A.D. Proc. 5th Int. Congr. Photosynth.: Halkidiki, Greece, 1981; in press.
25. Harris, N.; Dodge, A.D. Planta 1972, 104, 210.
26. Zweig, G.; Shavit, N.; Avron, M. Biochim. Biophys. Acta 1965, 109, 332.
27. Dodge, A.D. Endeavour 1971, 30, 130.
28. Mees, G.C. Ann. Appl. Biol. 1960, 48, 601.
29. Merkle, M.G.; Leinweber, C.L.; Bovey, R.W. Plant Physiol. 1965, 40, 832.

30. Rensen, J.J.S. van. Physiol. Plant 1975, 33, 42.
31. Lengfelder, E.; Elstner, E.F. Hoppe-Seylers Z. Physiol. Chem. 1978, 359, 751.
32. Youngman, R.J.; Dodge, A.D. Z. Naturforsch. 1979, 34C, 1032.
33. Youngman, R.J.; Dodge, A.D.; Lengfelder, E.; Elstner, E.F. Experientia. 1979, 35, 1295.
34. Farrington, J.A.; Ebert, M.; Land, E.J.; Fletcher, K. Biochim. Biophys. Acta 1973, 314, 372.
35. Bors, W.; Saran, M.; Lengfelder, E.; Spöttl, R.; Michel, C. Curr. Top. Radiat. Res. 1974, 9, 247.
36. Halliwell, B. FEBS Lett. 1976, 72, 8.
37. Koppenol, W.H.; Butler, J.; Leeuwen, J.W. van. Photochem. Photobiol. 1978, 28, 655.
38. Davenport, H.E. Proc. R. Soc. B. 1963, 157, 332.
39. Farrington, J.A. Proc. Brit. Crop. Prot. Conf. 1976, 225.
40. Harris, N.; Dodge, A.D. Planta 1972, 104, 201.
41. Harvey, B.M.R.; Muldoon, J.; Harper, D.B. Pl. Cell Environ. 1978, 1, 203.
42. Youngman, R.J.; Dodge, A.D. Proc. 5th Int. Congr. Photosynth.: Halkidiki, Greece, 1981; in press.
43. Harvey, B.M.R.; Fraser, T.W. Pl. Cell. Environ. 1980, 3, 107.
44. Harper, D.B.; Harvey, B.M.R. Pl. Cell. Environ. 1978, 1, 211.

RECEIVED September 2, 1981.

5

Interaction of Herbicides with Cellular and Liposome Membranes

DONALD E. MORELAND, STEVEN C. HUBER, and WILLIAM P. NOVITZKY

North Carolina State University, U.S. Department of Agriculture, Agricultural Research Service, Departments of Crop Science and Botany, Raleigh, NC 27650

Many herbicides inhibit chloroplast electron transport by binding to a protein located on the reducing side of PS II. Some of the herbicides (chlorpropham, dinoseb, propanil, ioxynil), but not all (diuron, *s*-triazines, uracils), also interfere with photophosphorylation and mitochondrial electron transport and phosphorylation. In this study, propanil, dinoseb, ioxynil, and several carbanilates, but not diuron, affected the following responses, much like the uncoupler FCCP: (a) inhibited the light-dependent quenching of atebrin fluorescence in thylakoids; (b) inhibited valinomycin-induced swelling of all organelles; (c) increased the permeability of all organelle membranes to K^+ in the absence of an ionophore; and (d) increased the permeability of phosphatidyl choline liposomes to H^+. The structure/activity correlations were similar for all organelles, i.e., the responses were not membrane specific. Results obtained suggested that when herbicides partition into the lipid phases of organelle membranes, perturbations are produced that lead to alterations in "fluidity" and permeability to cations. The alterations may be responsible for uncoupling of ATP generation in mitochondria and chloroplasts, and inhibition of electron transport in mitochondria.

A large number of commercial herbicides interfere with electron transport and ATP production in isolated chloroplasts and mitochondria (1). These herbicides can be divided into two groups: electron transport inhibitors and inhibitory uncouplers (1, 2). The dimethylphenylureas, substituted uracils, *s*-triazines, and pyridazinones have been classified as electron transport inhibitors, whereas the alkylated dinitrophenols, acylanilides, halogenated benzonitriles, and *N*-phenylcarbamates have been classified as inhibitory uncouplers. All of the above-named herbicides inhibit chloroplast electron transport by binding reversibly to a protein(s) associated with the B complex (3, 4, 5). B is considered to function as the secondary acceptor of electrons from PS II. Binding affinity of the herbicidal inhibitors has been correlated with inhibition of PS II. Additional details of the binding responses of the herbicides and chemical models proposed to explain the interaction between inhibitors and the receptor target are provided in other contributions published herein (5, 6, 7).

The electron transport inhibitors do not directly affect photophosphorylation or interfere with mitochondrial electron transport and phosphorylation (1). However, the inhibitory uncouplers, in addition to interfering with electron transport in thylakoids, uncouple photophosphorylation and oxidative phosphorylation, and inhibit mitochondrial electron transport.

Whereas inhibition of chloroplast electron transport has been correlated with binding to a protein(s), the sites and mechanisms through which herbicides interfere with mitochondrial and chloroplast mediated phosphorylations remain to be identified. When lipophilic herbicides partition into the lipid phases of membranes, they could perturb lipid-lipid, lipid-protein, and protein-protein interactions that are required for membrane functions such as electron transport, ATP formation, and active transport. Evidence for general membrane perturbations caused by chlorpropham, 2,6-dinitroanilines, perfluidone, and certain phenylureas have been reported previously (8-11).

The objectives of the studies reported here were to compare the action of the compounds identified in Table I (a) on chloroplast and mitochondrial electron transport and phosphorylation, and (b) on the "fluidity" and permeability to K^+ and H^+ of chloroplast, mitochondrial, and liposome membranes. For comparative purposes, dinoseb, ioxynil, propanil, and 3-CIPC (chlorpropham) were included to represent the inhibitory uncouplers. Diuron represented the electron transport inhibitors, and FCCP was included as a reference uncoupler. Three other carbanilates (3-CHPC; 2,3-DCIPC; and 3,4-DCIPC) were selected to determine the effect of replacement of the isopropyl side chain with a hexyl moiety and the effect of dichlorination of the phenyl ring in the 2,3- and 3,4-positions. In the phenylureas and acylanilides, increasing the length of the side chain has been associated with increased inhibitory activity against the Hill reaction (12). Also, in the phenylureas, phenylcarbamates, and acylanilides, dichlorination in the 3,4-ring positions is more inhibitory to chloroplast electron transport than monochlorination in either position. Additionally, chlorination in an ortho position has been associated with decreased inhibitory activity (12).

TABLE I
Common names or designations and chemical names of compounds studied.

Common name or designation	Chemical name
FCCP	carbonylcyanide *p*-trifluromethoxyphenylhydrazone
Dinoseb	2-*sec*-butyl-4,6-dinitrophenol
Diuron	3-(3,4-dichlorophenyl)-1,1-dimethylurea
Ioxynil	4-hydroxy-3,5-diiodobenzonitrile
Propanil	3',4'-dichloropropionanilide
3-CIPC	isopropyl *m*-chlorocarbanilate
3-CHPC	hexyl *m*-chlorocarbanilate
2,3-DCIPC	isopropyl 2,3-dichlorocarbanilate
3,4-DCIPC	isopropyl 3,4-dichlorocarbanilate

Materials and Methods

Chloroplasts. Intact chloroplasts were isolated from freshly harvested growth chamber-grown spinach (*Spinacia oleracea* L.) as described by Lilley and Walker (13). Thylakoids were prepared by the method of Armond *et al.* (14). Chlorophyll concentrations were determined by the method of MacKinney (15). Reaction cuvettes were illuminated at 25 C with a photon fluence rate of 7.5×10^{-4} mol/m^2·s (PAR). Reduction of ferricyanide with water as the oxidant was measured spectrophotometrically at 420 nm in a reaction medium (5.0 ml volume) that contained 0.1 M sorbitol, 50 mM tricine-NaOH (pH 8.0), 0.4 mM KH_2PO_4, 5 mM $MgCl_2$, 10 mM NaCl, 0.5 mM $K_3Fe(CN)_6$, 1 mM ADP, and thylakoids (100 µg chlorophyll). Esterification of inorganic phosphate was measured by the procedure of Lanzetta *et al.* (16). Effects on the light dependent quenching of atebrin were measured in a reaction medium (2.0 ml volume) that contained 1.0 µM atebrin, 0.1 M sorbitol, 10 mM tricine-NaOH (pH 7.8), 10 mM NaCl, and thylakoids (20 µg chlorophyll). The thylakoids were illuminated with red actinic light (Corning 2-64 filter). The atebrin was excited with 366 nm light and emission was observed at 505 nm.

Effects imposed on the efflux of K^+ were measured by suspending intact chloroplasts (100 µg chlorophyll) in a reaction medium (1.0 ml volume) that contained 0.4 M sorbitol, and 10 mM Hepes-NaOH (pH 7.1). After incubation with test chemicals for 2 min, the reaction mixtures were centrifuged to pellet the chloroplasts. K^+ content of the supernatants was measured by flame photometry.

Mitochondria. Mitochondria were prepared from 3-day-old dark-grown mung bean (*Phaseolus aureus* Roxb.) hypocotyls. The isolation procedure, measurements of oxygen utilization, and effects of the test compounds on respiratory states were conducted as described previously (9). Following the terminology of Chance and Williams (17), the ADP-stimulated rate of respiration will be referred to as state 3 and ADP-limited respiration as state 4. The mung bean mitochondria had respiratory control (state 3/state 4) ratios that averaged 3.9, 3.6, and 2.2; and calculated ADP/O ratios that averaged 2.3, 1.3, and 1.5, for the oxidation of malate, NADH, and succinate, respectively.

Osmotic Swelling. Changes in the osmotic stability of the organelles were monitored spectrophotometrically at 550 nm for chloroplasts and thylakoids, and at 520 nm for mitochondria. The 2.0-ml reaction mixture contained 0.15 M KSCN or KCl and 10 mM Hepes-NaOH (pH 7.1). The initial absorbance of the reaction was adjusted to 0.8, 0.4, and 0.75 for chloroplasts (approximately 20 μg chlorophyll), thylakoids (approximately 20 μg chlorophyll), and mitochondria (approximately 0.4 mg protein), respectively. Induction of passive swelling was measured with the test compound being added 30 s after the introduction of the organelles. In studies with valinomycin (0.1 μM), the test compound was added 30 s prior to the introduction of the ionophore. Rates of swelling were calculated from the initial phase of absorbance decrease.

Liposomes. Liposomes were prepared by sonication from egg yolk phosphatidyl choline (Sigma type X-E) according to the method of Hinkle (18). The assay medium used to determine effects of herbicides and FCCP contained 0.2 ml liposomes in 1.8 ml of 0.3 M NaCl, 20 mM tris-HCl (pH 7.5), 5 mM Na-ascorbate, 80 μM ferrocene, and 80 μM tetraphenylboron. Ferricyanide reduction was measured spectrophotometrically at 420 nm and 25 C.

Test Chemicals. Stock solutions of the desired concentrations of test chemicals were prepared in acetone. The final concentration of the solvent was held constant at 1% (v/v) in all assays including the controls. Data presented for the several studies were averaged from determinations made with a minimum of three separate replications and isolations.

Results and Discussion

Shown in Table II are I_{50} values for inhibition of electron transport in spinach thylakoids (water to ferricyanide) and the associated phosphorylation reaction by the test compounds. Also included are I_{50} values for inhibition of the light-dependent quenching of atebrin fluorescence and the ratio obtained by dividing the I_{50} for the atebrin response by the I_{50} for inhibition of photophosphorylation. The relative order of inhibitory potency for some of the compounds for inhibition of ferricyanide reduction has been reported previously (12). For the herbicides,

TABLE II
Effects of FCCP, selected herbicides, and carbanilates on reactions mediated by spinach thylakoids.

Compound	Ferricyanide Reduction	Ferricyanide Phosphorylation	Atebrin fluor.	Atebrin/ phosph. ratio
	---------------- I_{50} (μM) ----------------			
FCCP	NE[a]	0.35	0.4	1
Dinoseb	7.5	4	23	6
Diuron	0.07	0.08	NE	--
Ioxynil	0.3	0.2	120	600
Propanil	0.7	0.6	200	333
3-CIPC	150	140	130	1
3-CHPC	62	15	28	2
2,3-DCIPC	130	125	NE	--
3,4-DCIPC	12	8	70	9

[a] NE = no, or minimal, effect with concentrations up to 400 μM.

except diuron (a pure electron transport inhibitor), the I_{50} for inhibition of the coupled phosphorylation was lower than the I_{50} for inhibition of ferricyanide reduction. A lower I_{50} value for inhibition of phosphorylation suggests that the compounds might be expressing an effect on the phosphorylation pathway that cannot be explained entirely by interference with electron transport. The herbicides, except for diuron, have been shown previously to inhibit cyclic photophosphorylation measured in the absence of oxygen (1), and the light-induced synthesis of ATP mediated by dithiothreitol and PMS (19). The latter reactions do not involve the entire electron transport chain and are insensitive to diuron.

An increase in lipophilicity of 3-CIPC with the replacement of a hexyl for the isopropyl side chain (3-CHPC versus 3-CIPC) resulted in increased inhibition of electron transport and an even greater inhibition of the coupled phosphorylation (Table II). Increased reductive inhibitory potency of 3,4-DCIPC over 3-CIPC also is shown.

The inhibitory uncouplers and the uncoupler FCCP, but not diuron and 2,3-DCIPC, also inhibited the light dependent quenching of atebrin fluorescence in spinach thylakoids (Table II). Inhibition of the light-dependent quenching of atebrin fluorescence has been attributed to dissipation of the energized state (Δ pH) of the thylakoid membrane, which is considered to provide the driving force for phosphorylation (20, 21). The interferences observed support the suggestion that the inhibitory uncoupler herbicides do act as uncouplers. Inhibition of light-induced atebrin fluorescence quenching by the carbanilate derivatives also was strengthened by an increase in side chain lipophilicity (3-CIPC versus 3-CHPC) and dichlorination (3-CIPC versus 3,4-DCIPC), but negated by ortho substitution (2,3-DCIPC). Of the phenolic herbicides, dinoseb was a good uncoupler, whereas ioxynil was a relatively poor uncoupler. The inhibitory uncouplers, but not the electron transport inhibitors, also stimulated electron transport through PS I by illuminated thylakoids (reduced DPIP as oxidant and methyl viologen as reductant), when electron flow through PS II was blocked with diuron (22). Stimulation of oxygen uptake (22) and inhibition of the atebrin-induced fluorescence quenching (Table II) occurred over the same molar concentration ranges. Neither reaction inovlved PS II.

The atebrin/phosphorylation ratio (Table II, last column) relates inhibition of photophosphorylation to dissipation of the postulated energized state (Δ pH) of the thylakoid membrane. A low ratio suggests that the two responses are correlated. However, the high ratios obtained for ioxynil and propanil suggest that these two herbicides act as electron transport inhibitors rather than as uncouplers.

Mitochondrial Responses. The herbicides referred to as inhibitory uncouplers were so named because at low molar concentrations they satisfy most, if not all, of the criteria established for uncouplers of oxidative phosphorylation. However, at higher molar concentrations, they also inhibit mitochondrial electron transport (1).

Uncoupling action (interference with ATP generation) has been demonstrated by stimulation of state 4 respiration for the oxidation of malate, succinate, and NADH; circumvention of oligomycin-inhibited state 3 respiration; and induction of ATPase activity (1). In addition, by using various substrates, partial reactions, and electron mediators, evidence has been presented that the herbicides inhibit malate oxidation and malate-PMS oxidoreductase by acting at or near complex I, succinate oxidation and succinate-PMS oxidoreductase by acting at or near complex II, exogenous-NADH oxidation by acting prior to the cytochrome chain, and cyanide-resistant respiration (alternate oxidase). That inhibition of malate, succinate, and exogenous-NADH oxidation is not caused by interference at a common site, shared by the 3 substrates, is indicated by the widely differing I_{50} values that are obtained (1, 8, 9, 10). The pure electron transport inhibitors of chloroplast electron transport have only a marginal, if any, effect on mitochondrial responses.

For the most part, with the compounds included in this study, maximum uncoupling activity (stimulation of state 4 respiration) was obtained for the oxidation of succinate and inhibition of malate state 3 respiration was most sensitive. Therefore, only data for state 4 stimulation for the oxidation of succinate and

TABLE III
Effects of FCCP, selected herbicides, and carbanilates on reactions mediated by mung bean mitochondria.

Compound	Succinate state 4 respiration (Concn.)	(Stim.)	Malate state 3 respiration (Inhib.)
	μM	%	I_{50}(μM)
FCCP	0.1	108	50
Dinoseb	1	92	55
Diuron	NE[a]	NE	410
Ioxynil	10	90	150
Propanil	200	80	170
3-CIPC	200	71	100
3-CHPC	20	71	28
2,3-DCIPC	NE	NE	90
3,4-DCIPC	40	69	47

[a] NE = no, or marginal, effect with concentrations up to 400 μM.

inhibition of state 3 respiration for the oxidation of malate are presented in Table III. Only diuron and 2,3-DCIPC failed to stimulate state 4 respiration. None of the compounds was as effective an uncoupler as FCCP. Propanil and 3-CIPC showed maximum stimulatory activity at a relatively high molar concentration (200 μM). Replacement of the isopropyl group of CIPC for the hexyl group (3-CHPC) was associated with an increase in uncoupling activity as was the addition of a second chlorine in the 4-position (3,4-DCIPC) of the ring. For some of the herbicides, stimulation of state 4 respiration occurred at molar concentrations that also inhibited state 3 respiration, which can be attributed to an effect on electron transport. Hence, the effect on electron transport may have masked the full expression of uncoupling activity.

All of the compounds inhibited state 3 respiration with malate as substrate, however, diuron was a relatively weak inhibitor. As in the expression of uncoupling action, the lengthening of the side chain (3-CHPC versus 3-CIPC) and the addition of a second chlorine to the phenyl ring in the para position (3,4-DCIPC versus 3-CIPC) was associated with enhanced activity. The 2,3-DCIPC derivative was slightly less active than 3-CIPC, the reference carbanilate.

Whereas inhibition of chloroplast electron transport has been correlated with binding to a protein(s), the mechanisms for interference with photophosphorylation and the mitochondrial responses remain to be identified. The herbicides could act independently with several components of the mitochondrial electron transport and phosphorylation pathway, or they could be perturbing the membranes in such a way that multisite interference results.

Swelling Responses. Effects of the herbicides on the "fluidity" and permeability properties of membranes were measured. Alterations to the fluidity and permeability of organelle membranes can alter the osmotic properties of the membranes. Chloroplast, thylakoid, and mitochondrial membranes are known to be relatively impermeable to cations such as K^+ and H^+, but freely permeable to lipophilic anions such as SCN^- (23, 24). Hence, organelles are osmotically stable when suspended in isotonic KSCN. K^+ permeability, however, can be induced artificially by ionophores such as valinomycin (Figure 1A). Valinomycin is considered to form a lipid-soluble complex with K^+ and functions to transport K^+ across the membranes (25). An increase in internal K^+ will be accompanied by the diffusion of SCN^-, the counter ion, across the membrane to maintain electroneutrality. An increase in internal solute concentration will result in an influx of water and the organelles will swell.

Shown in the lowest traces of Figure 1B, C, and D is the rate and magnitude of swelling obtained when valinomycin was added to intact chloroplasts, thylakoids, and mitochondria, respectively. All of the test compounds except diuron, inhibited the rate and magnitude of valinomycin-induced swelling in the three organelles suspended in isotonic KSCN, as shown in Figure 1 for dinoseb. Dose/response curves were developed from the traces (Figure 1) and I_{50} values obtained from the curves are presented in Table IV. The effect was expressed at molar concentrations higher than those required to inhibit chloroplast electron transport and phosphorylation. However, the data demonstrate that inhibitory uncouplers can affect the properties of organelle membranes. The I_{50} values for inhibition of valinomycin-induced swelling in mitochondria, with the exception of FCCP, correlated closely with I_{50} values for inhibition of malate state 3 respiration (cf. Table III). Two of the carbanilates (3,4-DCIPC and 3-CHPC) were as inhibitory, or were better inhibitors, as indicated by the lower I_{50} values, than the standard uncoupler FCCP, in all three organelles.

The osmoticum used had a marked effect on the inhibitory potency of FCCP and the non-carbanilates (Table IV). For the non-carbanilate herbicides and FCCP, valinomycin-induced swelling of mitochondria suspended in isotonic KCl was inhibited at much lower concentrations of the herbicides than when the mitochondria were suspended in isotonic KSCN. For the carbanilates, the molar concentration for 50% inhibition of valinomycin-induced swelling of mitochondria suspended in KSCN was about twice the

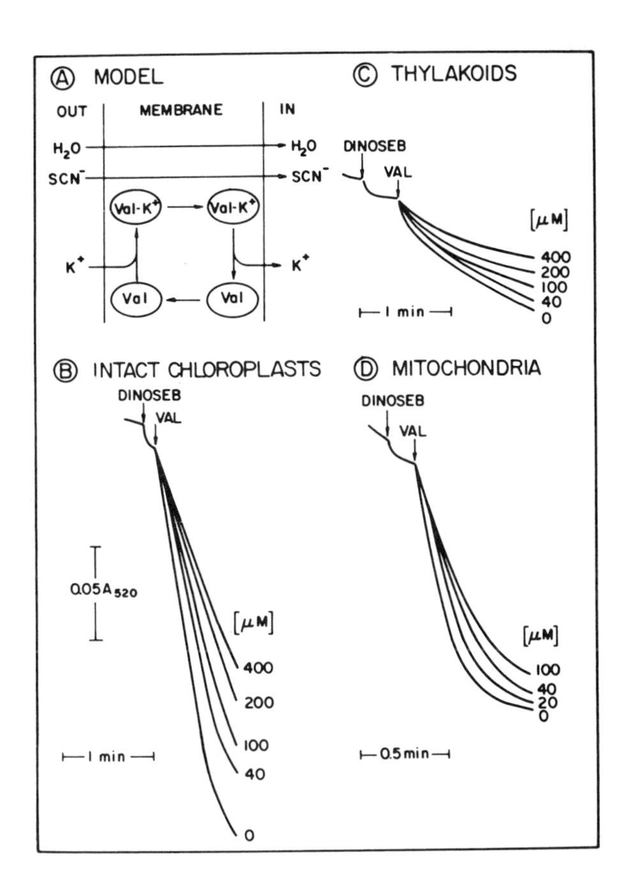

Figure 1. Representative traces of absorbance changes that show inhibition by dinoseb of valinomycin-induced swelling of intact spinach chloroplasts (B), spinach thylakoids (C), and mung bean mitochondria (D) suspended in isotonic KSCN. The model system is presented diagrammatically in A. Swelling was initiated by the addition of 0.1 μM *valinomycin (VAL).*

TABLE IV
Inhibition of valinomycin-induced swelling of intact spinach chloroplasts, spinach thylakoids, and mung bean mitochondria by FCCP, selected herbicides, and carbanilates.

Compound	Osmoticum			
	KSCN			KCl
	Chloroplasts	Thylakoids	Mitochondria	Mitochondria
		I_{50} (μM)		
FCCP	19	65	22	0.1
Dinoseb	220	170	90	0.4
Diuron	NE[a]	NE	NE	300
Ioxynil	310	240	330	2
Propanil	185	180	200	5
3-CIPC	180	135	185	100
3-CHPC	3	14	33	14
2,3-DCIPC	120	95	170	66
3,4-DCIPC	28	17	33	25

[a] NE = no, or minimal, effect with concentrations up to 400 μM.

concentration required to inhibit the response with mitochondria suspended in KCl.

Inhibition of valinomycin-induced swelling would occur if the test compounds interfered with movement of either K^+ or anion, or both. Because valinomycin is a mobile carrier of K^+, its rate of movement is dependent upon the fluidity of the membrane. One interpretation for the observed inhibitions is that the compounds, by partitioning into the organelle membranes, decrease the "fluidity" of the membrane, and, subsequently, the rate at which the valinomycin-K^+ complex moves across it. The greater sensitivity expressed with KCl as the osmoticum suggested that FCCP and the inhibitory uncouplers might also interfere, in some way, with Cl^- movement. The possibility exists that, in mung bean mitochondria, movement of Cl^- is carrier-mediated, whereas SCN^-, being a lipophilic anion, diffuses freely across the membrane. The herbicides may, in some manner, interfere with the endogenous transport mechanism and, therefore, cause greater inhibition of swelling in isotonic KCl.

Not all endogenous membrane transport systems are affected by the herbicides. For example, spontaneous mitochondrial swelling in isotonic solutions of ammonium phosphate or neutral amino acids, was affected only marginally by the compounds (data not shown). Swelling in these systems involves the endogenous Pi^-/OH^- antiporter (26) and amino acid porter (27), respectively. At this time, there is no ready explanation for the apparent inhibition of Cl^- transport by the non-carbanilates.

Effects on membrane permeability to ions. For the most part, the compounds induced chloroplasts, thylakoids, and mitochondria suspended in isotonic KSCN to swell in the absence of an ionophore. Typical results obtained with dinoseb are presented in Figure 2. As shown in the no-dinoseb control curves, some spontaneous swelling occurred at a slow rate. However, the addition of dinoseb greatly increased the rate and magnitude of the swelling response in all three organelles. In these experiments, as in the valinomycin studies, swelling would be expected to occur only if the permeability of the organelle membranes to K^+ was increased by the test compounds because the membranes are considered to be freely permeable to SCN^- (Figure 2A).

Dose/response curves were developed from traces such as those shown in Figure 2 for dinoseb. For comparative purposes, the concentration of compound required to induce an increase in the swelling rate of 0.02 A in 1 min relative to the no-herbicide controls is shown in Table V for chloroplasts, thylakoids, and mitochondria. Thylakoids did not swell as extensively as the other two organelles, consequently, the values are for a change of 0.01 A. The relative order of activity shown by the compounds is similar to that given in Table IV for inhibition of valinomycin-induced swelling in the three organelles.

Mitochondria suspended in isotonic KCl were also induced to swell by all of the compounds, except diuron (Table V). However, as opposed to inhibition of valinomycin-induced swelling (Table IV), the values obtained in isotonic KCl were slightly higher than those obtained in isotonic KSCN, with two exceptions (3-CHPC and 2,3-DCIPC). As in the valinomycin-induced system, in order for swelling to occur, both the cation and anion must cross the membranes. Because SCN^- is a permeant anion and its movement is controlled electrogenically, effects imposed by the compounds on the movement of K^+ would determine the occurrence and extent of swelling. Hence, with mitochondria suspended in KSCN, data reflect effects imposed on the movement of K^+. In mitochondria suspended in KCl, the compounds could affect movement of either or both K^+ and Cl^- because both are nonpermeant. If the compounds were increasing only the permeability of the mitochondrial membrane to K^+, then data obtained in KSCN and KCl would be expected to be similar. The higher values obtained in KCl relative to KSCN may reflect interference by the compounds with Cl^- movement.

As for inhibition of valinomycin-induced swelling, some of the carbanilates (3-CHPC and 2,3-DCIPC) seem to be acting differently in that values obtained in isotonic KCl were lower than those obtained in isotonic KSCN.

The permeability of chloroplast membranes to endogenous K^+ was also increased by FCCP and the inhibitory uncouplers, but not

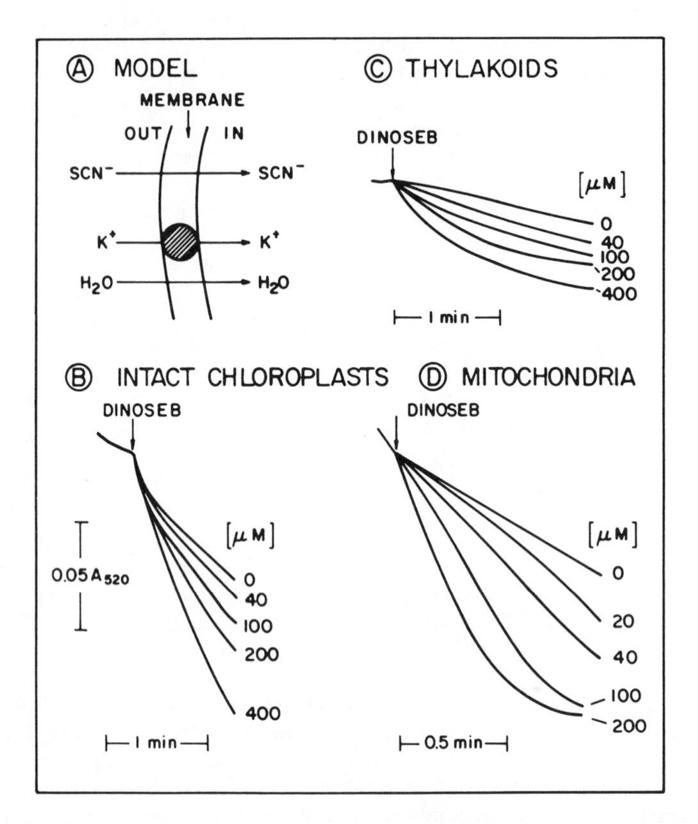

Figure 2. Representative traces of absorbance changes that show induction of passive swelling by dinoseb of intact spinach chloroplasts (B), spinach thylakoids (C), and mung bean mitochondria (D) suspended in isotonic KSCN. The model system is presented diagrammatically in A.

TABLE V
Induction of passive swelling of intact spinach chloroplasts, spinach thylakoids, and mung bean mitochondria by FCCP, selected herbicides, and carbanilates.

Compound	Osmoticum			
	KSCN			KCl
	Chloroplasts[a]	Thylakoids[b]	Mitochondria[a]	Mitochondria[a]
	I_{50} (μM)			
FCCP	110	31	9	30
Dinoseb	200	230	31	110
Diuron	NE[c]	NE	230	NE
Ioxynil	330	200	45	110
Propanil	160	200	59	220
3-CIPC	110	110	105	190
3-CHPC	14	14	34	24
2,3-DCIPC	71	NE	180	89
3,4-DCIPC	15	57	18	38

[a] Concentration (μM) required to induce an increase in the swelling rate of 0.02 A in 1 min, relative to the no-herbicide controls.

[b] Concentration (μM) required to induce an increase in the swelling rate of 0.01 A in 1 min, relative to the no-herbicide controls.

[c] NE = no, or minimal, effect with concentrations up to 400 μM.

the electron transport inhibitors (Table VI). The chloroplasts used in these studies contained about 250 nmoles stromal K^+/mg chlorophyll. A slow efflux of K^+ was induced by acetone (1%) used as a solvent for the test compounds. Data shown in Table VI have been corrected for the no-herbicide control efflux (40 nmoles K^+/mg chlorophyll·min). Diuron did not increase the efflux of K^+ significantly above that induced by acetone. Ioxynil, 3-CIPC, dinoseb, and 2,3-DCIPC induced significant K^+ efflux over a similar molar concentration range. Propanil had the weakest effect on K^+ efflux. The most active compounds were 3-CHPC, which initiated efflux at concentrations below 2 μM, and 3,4-DCIPC, which was a slightly better inducer than FCCP. The results directly support the postulate that the herbicides increased membrane permeability to K^+.

The inhibitory uncouplers, but not diuron and 2,3-DCIPC, also altered the permeability to protons of artificial, purely lipoidal, liposome membranes. In the system (Figure 3A), electrons flow

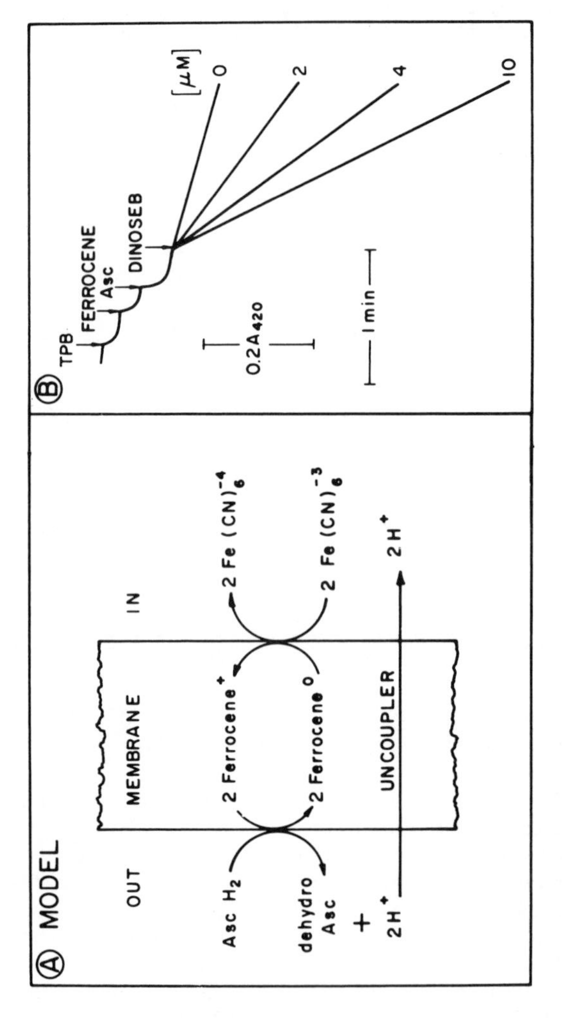

Figure 3. Effects of dinoseb on increasing the rate of reduction of ferricyanide, included within egg yolk phosphatidyl choline liposomes, by ferrocene (B). The model system is presented diagrammatically in A.

TABLE VI
Effects of FCCP, selected herbicides, and carbanilates on the efflux of potassium from intact spinach chloroplasts and the permeability of lecithin liposomes

Compound	K^+ efflux from chloroplasts[a]	H^+ permeability of liposomes[b]
	µM	
FCCP	22	.05
Dinoseb	110	1
Diuron	NE[c]	NE
Ioxynil	110	3
Propanil	290	120
3-CIPC	100	360
3-CHPC	6	49
2,3-DCIPC	120	NE
3,4-DCIPC	19	140

[a] Concentration (µM) required to induce efflux of K^+ at a rate of 100 nmoles/mg chlorophyll·2 min above the no-herbicide control rate.

[b] Concentration (µM) required to increase the rate of ferricyanide reduction to 80 nmoles/min. The no-herbicide control rate was about 40 nmoles/min.

[c] NE = no, or minimal, effect with concentrations up to 400 µM.

from ascorbate (outside) via ferrocene (in the liposome membrane) to ferricyanide (included within the liposome) in the presence of tetraphenylboron (18). Diffusion of H^+ across the liposome membrane is required to maintain electroneutrality.

Dose/response curves were developed from traces such as shown in Figure 3B for dinoseb. For comparative purposes, the concentration of compound required to increase the rate of ferricyanide reduction to twice that of the no-herbicide control rate are shown in the last column of Table VI. Uncouplers such as FCCP accelerate the rate of ferricyanide reduction, presumably by shuttling protons across the membrane in response to the electrical potential generated by the reduction of ferricyanide by ferrocene (28). In this study, FCCP was the most effective compound. The two phenolic herbicides (dinoseb and ioxynil) were more active than propanil and chlorpropham. Among the carbanilates, 3-CHPC and 3,4-DCIPC were more active than the parent compound (3-CIPC), whereas 2,3-DCIPC was essentially inactive. These results demonstrate that the inhibitory uncoupler herbicides, like the uncoupler FCCP, can increase the permeability of an artificial, purely lipoidal membrane to H^+.

In the liposome experiments as reported here, the no-herbicide control rate was affected by the relative concentration of ferrocene and tetraphenyl boron in the reaction mixture. Additionally, responses induced by the compounds varied quantitatively with the source and purity of the phosphatidyl choline from which the liposomes were prepared. However, the relative order of activity of the compounds remained constant. Data were reported for the egg yolk preparation because it provided a better evaluation of the relative activities of the carbanilates.

Conclusions

From the studies and results reported herein, the following general conclusions and extrapolations can be made:

a. The structure/activity correlations apply to chloroplasts, thylakoids, mitochondria, and liposomes, i.e., the responses observed were not membrane specific.

b. The inhibition of valinomycin-induced swelling (attributed to decreased membrane "fluidity") and induction of passive swelling (attributed to increased permeability of the membranes to K^+) are not organelle specific responses.

c. The limited structure/activity studies conducted with the carbanilates suggest that affinity for binding to the B protein complex correlates with increased uncoupling activity in PS I assays, increased uncoupling activity in mitochondria, enhanced membrane perturbations, enhanced permeability of chloroplasts to K^+, and enhanced permeability of liposomes to H^+.

d. The action on photophosphorylation may be separate and independent from binding to the B protein complex.

e. The inhibitory uncouplers may conceivably perturb all cellular membranes (plasmalemma, tonoplast, nuclear, and endoplasmic reticulum in addition to the chloroplast and mitochondrial membranes). However, marker systems that can be monitored readily which reflect the perturbations remain to be identified. This physiochemical interaction of the inhibitors with lipids can be expected to alter the many transport, biosynthetic, and regulatory activities associated with cellular membranes and could be involved in the expression of phytotoxicity.

The relation between membrane perturbations and the expression of phytotoxicity remains to be identified. In some of the experiments reported, concentrations in the 100 to 200 μM range were required to produce I_{50} responses. However, in many of these studies, initial effects could be detected at concentrations of 1 to 10 μM for many of the compounds. At this time, it is not possible to determine the impact of minor alterations to the permeability and "fluidity" of organelle membranes on the physiological status of an organism. Conceivably, small changes, coupled with a reduction in the availability of chloroplast and mitochondrially generated ATP, could have a significant and drastic effect over a time span of several hours or days.

Acknowledgements

This was a cooperative investigation of the North Carolina Agricultural Research Service and the United States Department of Agriculture, Agricultural Research Service, Raleigh, N.C. Paper No. 7079 of the Journal Series of the North Carolina Agricultural Research Service, Raleigh, N.C. The study was supported in part by Public Health Service Grant ES 00044. Appreciation is extended to F. S. Farmer for technical assistance with the mitochondrial phases of the study.

Literature Cited

1. Moreland, D. E. Annu. Rev. Plant Physiol. 1980, 31, 597-638.
2. Moreland, D. E.; Hilton, J. L. in "Herbicides: Physiology, Biochemistry, Ecology", Vol. 1; Audus, L. J., Ed.; Academic Press: London, 1976; pp. 493-523.
3. Tischer, W.; Strotman, H. Biochim. Biophys. Acta 1977, 460, 113-25.
4. Pfister, K.; Arntzen, C. J. Z. Naturforsch. 1977, 34c, 996-1009.
5. Steinback, K. E.; Pfister, K.; Arntzen, C. J. Chapter 3 in this book.
6. Shipman, L. L. Chapter 2 in this book.
7. Van Assche, C. J.; Carles, P. M. Chapter 1 in this book.
8. Moreland, D. E.; Huber, S. C. in "Plant Mitochondria"; Ducet, G.; Lance, C., Eds.; Elsevier/North Holland Biomedical Press: Amsterdam, 1978; pp. 191-8.
9. Moreland, D. E.; Huber, S. C. Pestic. Biochem. Physiol. 1979, 11, 247-57.
10. Moreland, D. E. Pestic. Biochem. Physiol. 1981, 15, 21-31.
11. Ducruet, J. M.; Gauvrit, C. Weed Res. 1978, 18, 327-34.
12. Moreland, D. E. in "Progress in Photosynthesis Research", Vol III; Metzner, H., Ed.; Int. Union Biol. Sci.: Tübingen, 1969; pp. 1693-1711.
13. Lilley, P. McC; Walker, D. A. Biochim. Biophys. Acta 1974, 368, 269-78.
14. Armond, P. A.; Arntzen, C. J.; Briantais, J.-M.; Vernotte, C. Arch. Biochem. Biophys. 1976, 175, 54-63.
15. MacKinney, G. J. Biol. Chem. 1941, 140, 315-22.
16. Lanzetta, P. A.; Alvarez, L. J.; Reinach, P. S.; Candia, O.A. Anal. Biochem. 1979, 100, 95-7.
17. Chance, B.; Williams, G. R. J. Biol. Chem. 1955, 217, 409-27.
18. Hinkle, P. Biochem. Biophys. Res. Commun. 1970, 41, 1375-81.
19. Alsop, W. R.; Moreland, D. E. Pestic. Biochem. Physiol. 1975, 5, 163-70.

20. Good, N. E. in "Encyclopedia of Plant Physiology, New Series", Vol. 5; Trebst, A.; Avron, M., Eds.; Springer-Verlag: Berlin, 1977; pp. 429-36.
21. Fiolet, J. T. W.; Bakker, E. P.; Van Dam, K. Biochim. Biophys. Acta 1974, 368, 432-45.
22. Moreland, D. E.; Huber, S. C.; Novitzky, W. P. Proc. 5th Int. Congr. Photosynth.: Halkidiki, Greece, 1981; in press.
23. Chappel, J. B.; Crofts, A. R. in "Regulation of Metabolic Processes in Mitochondria"; Tager, J. M.; Pappa, S.; Quagliariello, E.; Slater, E. C., Eds.; Elsevier: Amsterdam, 1966; pp. 293-316.
24. Jagendorf, A. T. in "Bioenergetics of Photosynthesis"; Govindjee, Ed.; Academic Press: New York, 1975; pp. 413-92.
25. Pressman, B. C. Annu. Rev. Biochem. 1976, 45, 501-29.
26. Huber, S. C.; Moreland, D. E. Plant Physiol. 1979, 64, 115-9.
27. Cavalieri, A. J.; Huang, A. H. C. Plant Physiol. 1980, 66, 588-91.
28. Bakker, E. P.; Van Den Heuvel, E. J.; Wiechmann, A. H. C.; Van Dam, K. Biochim. Biophys. Acta 1973, 292, 78-87.

RECEIVED September 21, 1981.

6

Effects of Herbicides on the Lipid Composition of Plant Membranes

J. B. ST. JOHN

U.S. Department of Agriculture, Beltsville Agricultural Research Center, Beltsville, MD 20705

Substituted pyridazinone herbicides directly inhibit photosystem II, chloroplast pigment biosynthesis, and membrane lipid biosynthesis. Depending on the substitution, pyridazinones can: specifically inhibit the synthesis of linolenic acid in galactolipids and phospholipids; preferentially alter the fatty acid composition of monogalactosyl diglycerides compared with digalactosyl diglycerides; and cause a build-up of saturated fatty acids in the chloroplast membranes. The differential responses of plant species and tissues to substituted pyridazinones suggest that control of linolenic acid biosynthesis may vary depending on plant species and even tissue. An interaction between triazine herbicides and the lipid composition of chloroplast membranes may influence sensitivity of weed biotypes to the triazine herbicides.

Substituted pyridazinone herbicides may have a multifunctional mode of action. Inhibitory action by the substituted pyridazinones has been demonstrated at three separate sites (1, 2, 3). The first site involves interference with photosynthetic functions (1). The second site involves interference with the accumulation of chloroplast pigments and the third site involves interference with the formation of chloroplast glycerolipids (2, 3). The molecular structures used most frequently to evaluate action at these sites are depicted in Figure 1.

Photosynthetic Electron Transport Inhibition

Numerous herbicides are known to inhibit photosystem II (PS II) - dependent electron transport. The values for inhibition by selected pyridazinones of electron transport in isolated barley chloroplasts are shown in Table I. The value for diuron [3-(3,4-dichlorophenyl)-1,1-dimethyl urea] may be used to relate these results to those in other literature.

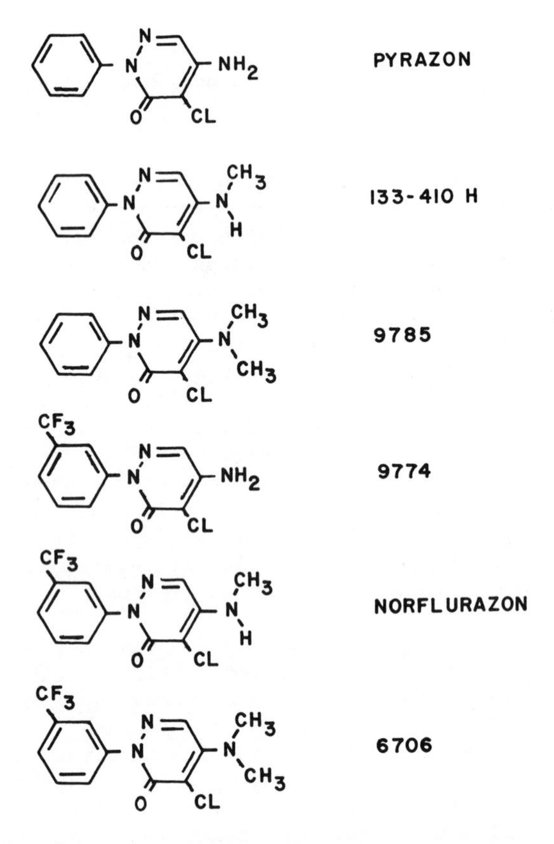

Figure 1. Structure of six substituted pyridazinones (2).

Table I
Herbicide inhibition of ferricyanide reduction by the Hill reaction in isolated chloroplasts (1).

Herbicide	Molar concn. for 50% inhibition
Pyrazon	7.0×10^{-6}
San 9785 (BASF 13 338)	1.4×10^{-5}
San 9774	1.4×10^{-5}
San 6706	4.4×10^{-6}
Diuron	1.4×10^{-7}

Inhibition of PS II is thought to occur at the level of a proteinaceous component located between plastiquinones Q and B, the primary and secondary acceptors of electrons from PS II. The competitive binding experiments of Tischer and Strotmann (4) suggest that the phenylureas, biscarbamates, triazines, triazinones, and pyridazinones inhibit electron transport by interaction with the same component of PS II. Action at this site seemed to account for the phytotoxicity of pyrazon [5-amino-4-chloro-2-phenyl-3(2*H*)-pyridazinone]. In addition to action at this site, compounds with molecular substitutions onto the structure of pyrazon (Figure 1) also interfere with the formation of chloroplast membrane lipids, namely the chlorophylls, carotenoids, and glycerolipids.

Chloroplast Pigment Inhibition

The chlorophylls and carotenoids are the pigmented lipids associated with chloroplast membranes. Substituted pyridazinones reduce the accumulation of these pigments (Table II).

Table II
Effect of substituted pyridazinones on chloroplast pigment accumulation in 4-day-old wheat shoots (2).

Treatment	Chlorophyll	Carotenoids
	% of control	
Pyrazon[a]	93	88
San 133-410H	60	60
San 9785 (BASF 13 338)	102	98
San 9774	45	57
Norflurazon (San 9789)	4	<1
San 6706	1	<1

[a] All chemicals were evaluated at a concentration of 1×10^{-4} M.

Pyrazon, the least substituted pyridazinone evaluated, did not alter pigment accumulation. Adding a trifluoromethyl group to the phenyl ring (San 9774) or monomethyl substitution of the amine (San 133-410H) resulted in almost 50% inhibition of pigment accumulation. Both substitutions (norflurazon) were required for maximum effectiveness. Dimethyl substitution of the amine (San 9785 = BASF 13 338) did not interfere with pigment accumulation. The action of San 6706 (trifluoromethyl substitution on the phenyl ring and dimethyl substitution on the amine) on pigment accumulation is more like that of San 9785. Thus, the data are consistent with the hypothesis that San 6706 is metabolized to norflurazon before exerting its effect on pigment accumulation in wheat shoots.

Photooxidation of the chlorophylls as a consequence of the absence of carotenoids, which have a protective function, is hypothesized to cause the bleaching effect in some systems (5, 6). However, simultaneous inhibition of carotenoid and chlorophyll accumulation has been demonstrated in other systems (7).

Numerous groups have used the pyridazinone action at this site as a tool for measurement of phytochrome *in vivo* in light grown seedlings (8, 9). Interference by chlorophylls and carotenoids had previously prevented *in vivo* measurement of phytochrome.

Pyridazinone Action on Galactolipids

Chloroplast membranes differ from most other membranes in that the major portion of the non-pigmented lipids of the chloroplast are the linolenic acid-rich galactolipids, with sulfolipids and phospholipids present as the minor lipid constituents. The third site of action affected by pyridazinones in wheat shoots is the formation of galactolipids (2, 3). The data in Table III summarize the effects of substituted pyridazinones on the relative fatty acid composition of wheat shoots.

Table III
Effect of substituted pyridazinones on relative fatty acid composition of galactolipids from wheat shoots (3).

Compound (0.1 mM)	Monogalactosyl-diglyceride		Digalactosyl-diglyceride	
	ΣS/U[a]	18:2/18:3[b]	ΣS/U[a]	18:2/18:3[b]
Control	0.22	0.32	0.27	0.17
Pyrazon	0.29	0.42	0.38	0.25
San 133-410 H	0.87	0.93	0.71	0.50
San 9785 (BASF 13 338)	0.35	3.82	0.39	3.32
San 9774	1.50	2.30	0.53	0.33
Norflurazon (San 9789)	1.21	1.67	0.81	1.80
San 6706	0.97	1.70	0.82	1.40

[a] Total saturated/total unsaturated fatty acids
[b] Linoleic (18:2) to linolenic (18:3) acid

Pyrazon, which did not affect pigment accumulation, did not affect the fatty acid composition of either galactolipid species. San 9774 preferentially altered the fatty acid composition of monogalactosyl digylceride (MGDG) compared to digalactosyl diglyceride (DGDG). Alterations of MGDG included a shift toward a relatively more saturated fatty acid composition as well as an increase in the relative proportion of linoleic to linolenic acid. San 9785 (BASF 13 338) altered only the relative proportion of linoleic to linolenic acid in both MGDG and DGDG. Norflurazon, San 6706, and San 133-410 H shifted the ratio of saturated to unsaturated fatty acids and of linoleic to linolenic acid in both classes of galactolipids. The action of San 6706 again more closely resembles that of norflurazon.

Separation of Modes of Action in Wheat Shoots

With actions at multiple sites, the possibility exists that actions at some sites may arise as secondary consequences of actions at other sites. The collective data argue that the multiple actions displayed by the pyridazinones are independent.

Pyrazon inhibited PS II (Table I) but did not interfere with the accumulation of chloroplast pigments (Table II) or influence the composition of glycerolipids (Table III). Therefore, even though the other substituted pyridazinones also inhibit PS II, their actions on chloroplast pigments and/or glycerolipids are not likely to result from PS II inhibitions. Further evidence that inhibition of PS II does not result in membrane lipid changes is presented in Table IV.

Table IV

Relative fatty-acid composition of polar lipids from light and dark grown wheat shoots in the presence of 1 x 10^{-4} M diuron (2).

Treatment		Relative polar lipid fatty acid composition: $\Sigma S/U$[a] (ratio)	Relative polar lipid fatty acid composition: 18:2/18:3[b] (ratio)
Dark	Control	0.33	1.48
	Diuron	0.33	1.55
Light	Control	0.23	0.51
	Diuron	0.23	0.47

[a] Ratio of total saturated to total unsaturated fatty acids

[b] Ratio of linoleic (18:2) to linolenic (18:3) acid

Diuron is a particularly potent inhibitor of the Hill reaction (Table I), but it does not significantly alter membrane lipid composition. The possibility that the effects of the pyridazinones on membrane lipids result from photooxidation as a secondary consequence of the carotenoid inhibition is not likely. Wheat shoots treated with BASF 13 338 have normal levels of carotenoids and chlorophylls (Table II), but linolenic acid levels are severely depressed (Table III). Thus, inhibition at the lipid site can occur without inhibition at the pigment site. Also, BASF 13 338 preferentially reduces linolenic acid, whereas the double bonds of all the unsaturated fatty acids would be susceptible to photooxidation. San 9774 preferentially alters the fatty acid composition of MGDG compared to DGDG (Table III). The preferential alteration of a single class of membrane lipids is inconsistent with the photooxidation hypothesis.

Frosch et al. (10) also concluded that the pyridazinone effect on the lipid composition of the plastid membrane did not result from photooxidation. They demonstrated that both BASF 13 338 and San 9789 caused changes in membrane lipid composition, but neither compound structurally altered the plastid membrane system, visualized with the electron microscope, under conditions excluding photodestruction of chlorophyll. When treated seedlings were transferred to white light, the thylakoids and plastid ribosomes disappeared only in the San 9789-treated seedlings. The effectiveness of the light in causing structural disturbances correlated with the effectiveness of light in mediating the photodestruction of chlorophyll. The structural disintegration was not related to the effect of the pyridazinones on the lipid composition of the plastid membrane since structural changes did not occur in the presence of BASF 13 338 even though the lipid compositional effects had occurred.

The lack of involvement of photosynthetic inhibitions in lipid inhibitions by pyridazinones is also supported by results in non-photosynthetic tissue (11, 12). The phospholipids predominate in root membranes. The data in Table V show that BASF 13 338 does not influence the distribution of lipid between the various classes of phospholipids, but BASF 13 338 specifically decreases the linolenic acid content of phosphatidylcholine and phosphatidylethanolamine.

Pyridazinone effects on phospholipids also provide information relative to phospholipid biosynthesis. The similarity in fatty acid composition of PC and PE in the absence of BASF 13 338 and their equivalent response to treatment suggest that the biosynthesis of phospholipids in wheat roots is not specific with respect to acyl chains.

Table V
Phospholipid composition for wheat roots (12).

Treatment	Phospholipid Fraction				
	PC[a]	PE	PG	PI	PA-PS
	% of total lipid phosphorus				
Control	51	34	5	7	4
BASF 13 338 (100 μm)	50	35	6	5	4

[a] PC = phosphatidyl choline; PE = phosphatidyl ethanolamine; PG = phosphatidyl glycerol; PI = phosphatidyl inositol; PA = phosphatidic acid; PS = phosphatidyl serine

Table VI
Fatty acid composition of phosphatidyl choline and phosphatidyl ethanolamine of wheat roots (12).

Treatment	Phospholipid	Fatty Acid[a]				
		16	18	18:1	18:2	18:3
		% by weight				
Control	PC	27	2	5	49	16
	PE	32	3	4	47	14
BASF 13 338	PC	25	2	5	63	5
	PE	27	5	7	56	4

[a] Fatty acid: 16 = palmitic; 18 = stearic; 18:1 = oleic; 18:2 = linoleic; 18:3 = linolenic acid

Differential Sensitivity to the Pyridazinones

Plant species differ in their susceptibility to the pyridazinones. Corn (Zea mays L.) was more sensitive than soybean (Glycine max L.) which was more sensitive than cotton (Gossypium hirsutim L.) based on visual observations of phytotoxicity (13). We have demonstrated differential susceptibility of the cereals to effects of BASF 13 338 on membrane-bound linolenic acid

(Figure 2) (14). We have also demonstrated that tissues within the same plant can vary in sensitivity to BASF 13 338 (Table VII) (15).

Table VII
Ratio of linoleic to linolenic acid in polar lipids of cotton seedlings (15).

Tissue	Growth Temperature	Treatment		
		Control	BASF 13 338 (10 μM)	Treated ÷ Control
			18:2/18:3	
Root tips	30°	2.67	4.04	1.51
	15°	1.16	2.56	2.21
Hypocotyls	30°	3.36	3.54	1.05
	15°	1.84	2.12	1.15

BASF 13 338 decreased the relative proportion of linolenic acid in the membranes of cotton root tips regardless of the growth temperature. In addition, BASF 13 338 blocked the low temperature-induced increase in linolenic acid in root tips. BASF 13 338 did not affect linolenic acid levels at 30° in hypocotyl tissue from the same cotton plants. The desaturation step only became sensitive to BASF 13 338 at low temperatures. This could indicate either a lower rate of desaturation in hypocotyls, a different control mechanism, or a difference in the desaturation enzymes between the two tissues.

The specificity of BASF 13 338 in inhibiting the desaturation of linoleic acid also demonstrates that the enzymes responsible for each desaturation of 18-carbon fatty acids are different. Furthermore, Khan *et al*. (16) have demonstrated a selective effect of San 6706 on the content of trans-Δ3-hexadecenoic acid (16:1) in phosphatidylglycerol from leaves of *Vicia faba*.

Lipid Composition and Triazine Resistance

Steinback *et al*.(17) have summarized the elegant work leading to their hypothesis that the receptor site for the PS II triazines is apparently an intrinsic membrane polypeptide of the 32- to 34-kilodalton size class. The photoaffinity label ^{14}C-azidoatrazine covalently binds with this polypeptide(s) in membranes of species susceptible to triazines. The thylakoids of resistant biotypes no longer bind the herbicide.

We compared the membrane lipid composition of chloroplasts isolated from species of common groundsel (*Senecio vulgaris* L.), lambsquarters (*Chenopodium album* L.), and pigweed (*Amaranthus hybridus* L.) (Table VIII) (18).

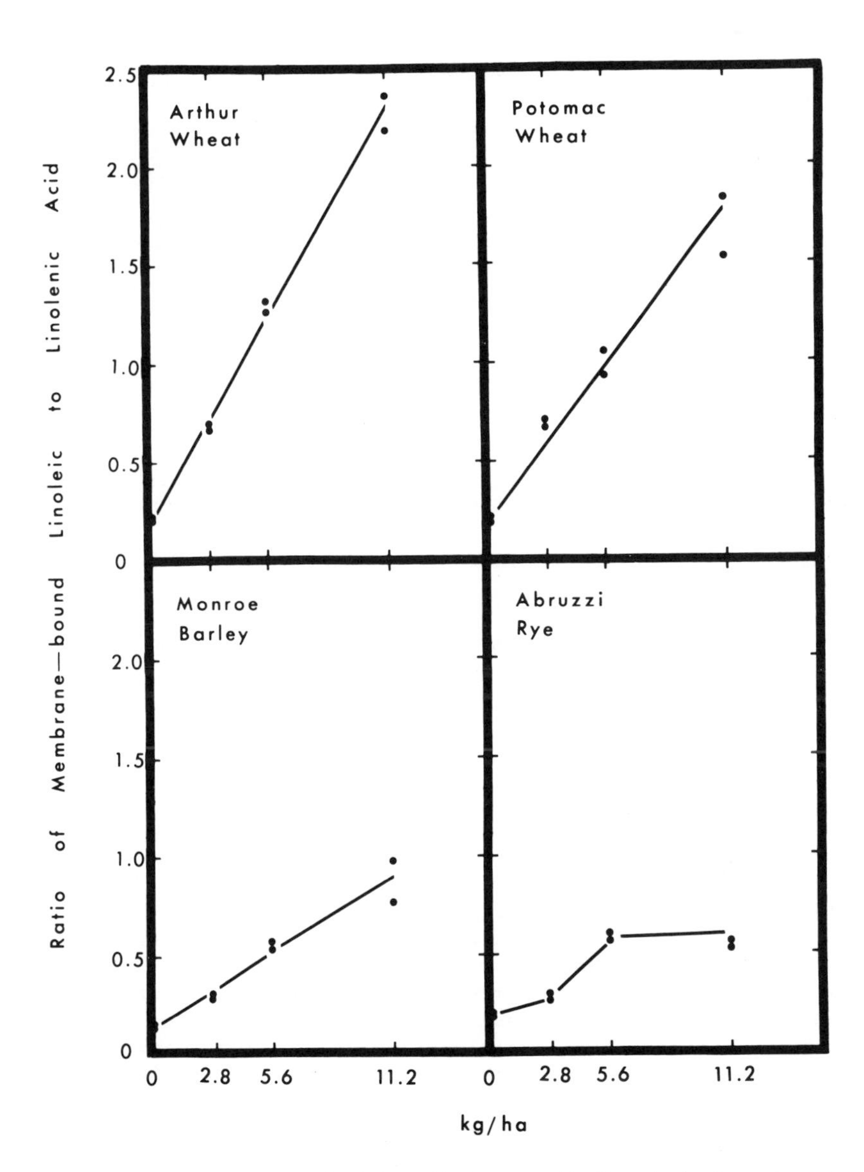

Figure 2. Effect of BASF 13 338 application rate (kg/ha, abscissa) on the ratio of linoleic to linolenic acid of field cultured wheat, barley, and rye (14).

Table VIII

Lipid composition of chloroplasts isolated from atrazine-resistant and atrazine-susceptible biotypes (18).

Plant Biotype	Resistant or Susceptible	Total Lipid	Polar Lipids			
			MGDG	DGDG	PE	PC
		% polar		% of total		
Groundsel	R	77.1[xxa]	52.9[x]	16.1[xx]	9.9[xx]	9.7
	S	68.6	48.7	28.5	5.5	11.5
Lambsquarter	R	82.8[xx]	54.4[xx]	21.3	10.7	8.8[xx]
	S	76.5	45.1	23.6	9.1	10.1
Redroot pigweed	R	78.0[xx]	55.6[xx]	20.1[xx]	7.9[xx]	10.0[xx]
	S	66.1	45.9	24.5	5.6	14.1
Average	R	79.3	54.3	19.2	9.5	9.5
	S	70.4	46.6	25.5	6.7	11.9

[a] Means are significantly different at the 10% (x) and 5% (xx) level, respectively, by a t test

A greater proportion of the total lipid from chloroplasts isolated from resistant species was found in the polar (membrane) lipid fraction. When this fraction was separated into individual lipid classes, chloroplasts from resistant biotypes contained higher proportions of MGDG and PE, and lower proportions of DGDG and PC than chloroplasts from susceptible species. There were no consistent trends for other classes of phospholipids. The proportional differences are a reflection of qualitative differences. However, a ratio of the resistant to susceptible value of greater than 1 in Table VIII indicates a quantitative increase in that lipid class and vice versa.

Differences were also noted in the fatty acid distribution in chloroplast lipids from atrazine-resistant and atrazine-susceptible biotypes (Table IX). Resistant chloroplast membranes contained lipids comparatively richer in unsaturated fatty acids with the exceptions of DGDG from all three biotypes and PE from common groundsel.

Table IX
Differences in distribution of lipid fatty acids from chloroplasts of atrazine-resistant and atrazine-susceptible biotypes (18).

Plant Biotype	Resistant or Susceptible	Unsaturation Ratio (18:2 + 18:3) / 16:0 Total	Polar	MGDG	DGDG	PE	PC
Groundsel	R	11.1[xa]	17.8[x]	13.7[xx]	1.8[xx]	1.5[xx]	3.5[xx]
	S	8.2	14.4	8.4	2.1	2.4	1.7
Lambsquarters	R	12.1[xx]	13.1[xx]	22.9[xx]	2.4	1.8[x]	2.9[xx]
	S	7.5	10.6	18.5	2.9	1.4	2.6
Redroot pigweed	R	9.7[xx]	14.6[xx]	16.4[xx]	2.3	2.0[xx]	2.8[xx]
	S	5.2	7.1	8.9	2.8	1.7	2.2
Average	R	10.9	15.2	17.7	2.2	1.8	3.1
	S	7.0	10.7	11.9	2.6	1.8	2.2

[a] Means are significantly different at the 10% (x) and 5% (xx) level, respectively, by a t test

Grenier et al. (19) have reported changes in the lipids and fatty acids of duckweed (Lemna minor L.) cultivated for 5, 10, or 15 days in mineral solutions containing sublethal concentrations of atrazine (0.05 to 0.75 ppm). All concentrations of atrazine used, independently of plant age, increased the total fatty acids, except for 5-day-old plants at 0.50 and 0.75 ppm atrazine where a decrease in total fatty acids was observed. Sublethal concentrations of atrazine increased the percentage of MGDG compared with total phospholipids and neutral lipids. Linolenic acid content increased while linoleic acid content decreased. MGDG was the main lipid involved in the linolenic acid increase observed in the total fatty acids. The changes that occur in lipids of duckweed cultured on sublethal dosages of atrazine are strikingly similar to the changes in lipids observed in resistant versus susceptible biotypes (Tables VIII and IX). Grenier et al. (19) suggested that the increase in linolenic acid and in MGDG in the presence of sublethal concentrations of atrazine were an indication that these treatments maintained intact and fully functional chloroplast membranes.

Heise and Harnischfeger (20) noted a positive correlation between the ratio of MGDG to total phospholipid and photophosphorylation efficiency. The data in Table X show that the ratio of MGDG to total phospholipid was consistently higher in chloroplast membranes of resistant biotypes (18).

Table X
Ratio of MGDG to total phospholipids (18).

Species	Triazine sensitivity		
	Sensitive	Resistant	Ratio R/S
Pigweed	1.89	2.48	1.31
Lambsquarter	2.15	2.38	1.10
Groundsel	1.28	1.91	1.50

Conard and Radosevich (21) and Holt *et al*. (22) have presented contrasting evidence on photosynthetic efficiency. They suggest that modification of the herbicide binding site which confers triazine resistance also makes photochemical electron transport much less efficient. The alteration resulted in a lowered capacity for net carbon fixation and lower quantum yields in whole plants of the resistant types of *Senecio vulgaris* L.

At this point it is not clear whether the lipid changes reported herein are directly related to the resistance trait. Nothing is known about the inheritance of the lipid changes while the resistance trait is known to be maternally inherited (23). It is generally accepted that lipid saturation is influenced by light intensity and growth temperature. The lipid changes might simply reflect differential response to light and temperature of sensitive and resistant species. On the other hand, if the lipid changes are directly related to the resistance trait, then knowledge of the interaction between the membrane lipids and proteins of the PS II complex could provide basic insights into factors controlling photosynthetic efficiency as well as triazine resistance.

Literature Cited

1. Hilton, J. L.; Scharen, A. L.; St. John, J. B.; Moreland, D. E.; Norris, K. H. *Weed Sci.* 1969, 17, 541-7.
2. St. John, J. B.; Hilton, J. L. *Weed Sci.* 1976, 24, 579-82.
3. St. John, J. B. *Plant Physiol.* 1976, 57, 38-40.
4. Tischer, W.; Strotmann, H. *Biochim. Biophys. Acta* 1977, 460, 113-25.
5. Bartels, P. G.; Hyde, A. *Plant Physiol.* 1970, 45, 807-10.
6. Bartels, P. G.; McCullough, C. *Biochem. Biophys. Res. Comm.* 1972, 48, 16-22.
7. Kunert, K. J.; Böger, P. *Z. Naturforsch.* 1979, 34c, 1047-51.
8. Jabben, M.; Dietzer, G. F. *Plant Physiol.* 1979, 63, 481-5.
9. Briggs, W. R.; Gorton, H. L. *Plant Physiol.* 1980, 66, 1024-6.
10. Forsch, S.; Jabben, M.; Bergfeld, R.; Kleinig, H.; Mohr, H. *Planta* 1979, 145, 497-505.
11. Willemot, C. *Plant Physiol.* 1977, 60, 1-4.

12. Ashworth, E. N.; Christiansen, M. N.; St. John, J. B.; Patterson, G. W. Plant Physiol. 1981, 67, 711-15.
13. Strang, R. H.; Rogers, R. L. J. Agri. Food Chem. 1974, 22, 1119-25.
14. St. John, J. B.; Christiansen, M. N.; Ashworth, E. N.; Gentner, W. A. Crop Sci. 1978, 19, 65-9.
15. St. John, J. B.; Christiansen, M. N. Plant Physiol. 1976, 57, 257-9.
16. Khan, M. U.; Lem, N. W.; Chandorkar, K. R.; Williams, J. B. Plant Physiol. 1979, 64, 300-5.
17. Steinback, K. E.; Pfister, K.; Arntzen, C. J. Chapter 3 in this book.
18. Pillai, P.; St. John, J. B. Plant Physiol. 1981, In Press.
19. Grenier, G.; Marier, J. P.; Beaumont, G. Can. J. Bot. 1979, 57, 1015-20.
20. Heise, K. P.; Harnischfeger, G. "Advances in the Biochemistry and Physiology of Plant Lipids"; Appelqvist, L. A.; Liljenberg, C., Ed.; Elsevier North Holland Biomedical Press: Amsterdam, 1979; p. 175-80.
21. Conard, S. G.; Radosevich, S. R. J. Appl. Ecol. 1979, 16, 171-7.
22. Holt, J. S.; Stemler, A. J.; Radosevich, S. R. Plant Physiol. 1981, 67, 744-8.
23. Gasquez, J.; Darmency, H.; Compoint, P. C. R. Acad. Sci. Paris 1981, 292, 847-9.

RECEIVED September 14, 1981.

7

Mode of Action of Herbicidal Bleaching

GERHARD SANDMANN and PETER BÖGER

Universität Konstanz, Lehrstuhl für Physiologie und Biochemie der Pflanzen, D-7750 Konstanz, Federal Republic of Germany

Data on herbicides are presented and reviewed, which allows the distinction between two different modes of bleaching. The first mode is caused by inhibited carotene biosynthesis exhibited by particular phenylpyridazinones, substituted phenylfuranones or amitrole. Decrease of carotenes leads to subsequent photodestruction of chlorophyll, peroxidation of other membrane components, and decay of electron transport activity. The second mode, represented by p-nitrodiphenylethers, is associated with peroxidation of membrane-bound polyunsaturated fatty acids concurrently with the breakdown of carotenes, chlorophylls, and decay of photosynthetic electron transport. Short-chain hydrocarbon gases are reliable markers. The action of peroxidizing diphenylethers appears to be related to that of bipyridylium salts, although no light-induced oxygen uptake can be measured.

Peroxidation is started probably by a diphenylether radical originating through electron transport. In addition, peroxidative bleaching may occur when suboptimum conditions for photosynthetic electron transport prevail and light is absorbed by pigments, but cannot be channeled into chemical work. This may be caused either by degradative decay or direct inhibition of the electron transport system combined with higher light intensities and/or by lowering the level of protective carotenes.

Many modern herbicides act on the chloroplast, their primary target is the photosynthetic membrane (1,2). Four sites of primary herbicidal activity may be distinguished at the moment and the herbicides grouped accordingly, although many compounds belong to more than one group.

0097-6156/82/0181-0111$05.00/0

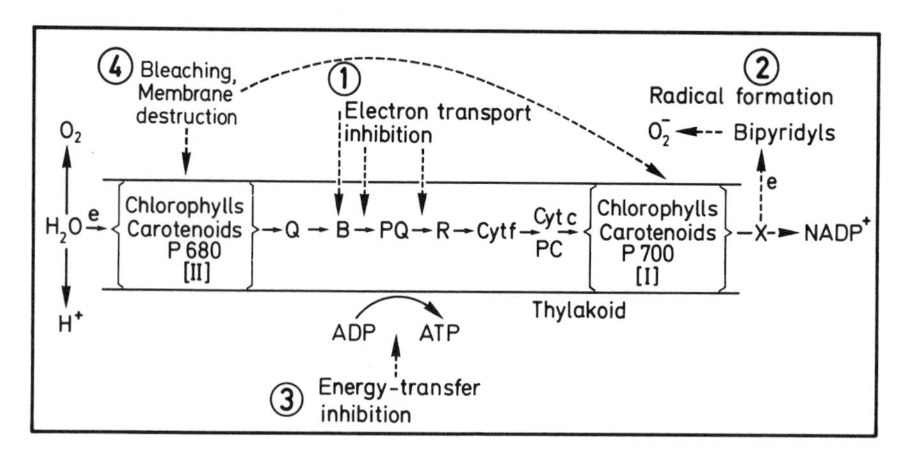

Figure 1. Sites of action of herbicides that affect the chloroplast at the photosynthetic electron transport units. Water is the electron donor, NADP the terminal acceptor. Light energy absorbed by chlorophylls and carotenoids of photosystems II and I is transferred to their respective reaction centers, P680 and P700. These are part of a chain of electron carriers like: Q, the primary acceptor for electrons donated by P680; B, the binding protein, which apparently functions as a Q/PQ reductase; and PQ, plastoquinone. R denotes a high-potential nonheme iron ("Rieske"-type) protein, which possibly forms a complex with cytochrome f. Cyt c and PC are (plastidic) cytochrome c-553 and plastocyanin. Presumably the R/Cyt f complex and B bind electron-transport inhibitor herbicides of group 1. Herbicides of group 2, mostly bipyridylium salts, accept electrons at X, which is possibly composed of membrane-bound nonheme iron centers. Group 3, at the moment represented by certain diphenyl ethers, affects the ATP synthetase ("energy-transfer inhibition"). Group 4 consists of "bleaching herbicides" of which two modes of action are known, namely inhibition of carotenoid biosynthesis and (lipid) peroxidation associated with membrane degradation.

The first are the electron-transport inhibitors such as ureas, symmetric triazines, triazinones, and certain phenylpyridazinones; also included are phenol compounds, bentazon or some diphenylethers (for chemical names of herbicides or formulas see legend of Figure 2). All have been shown to affect the B/plastoquinone region and the Rieske-protein cytochrome f/b_6 complex between the two photosystems (Figure 1). Considerable progress has been made to elucidate some binding sites in this region, particularly by using mutants resistant to certain inhibitors, or by trypsin treatment of isolated chloroplast material (see 3,4,5 for review). Undoubtedly, electron transport inhibitors are the most commercially important modern herbicides.

Group (2) are the bipyridylium salts which, according to their low midpoint-redox potential, may accept electrons at the reducing side of photosystem I (6). The bipyridylium radicals thus formed give rise to activated oxygen species, that may initiate breakdown of membrane components by a peroxidative process (7,8). Membrane destruction and the rapid decrease mainly of photosystem-II redox activity are thought to be the early and decisive herbicidal effect followed by pigment bleaching (9, see also for further refs.).

Group (3) is a rather small one comprising certain p-nitrophenylethers of which nitrofen is the most prominent. As we have shown (10) nitrofen has a strong inhibitory influence upon the ATP synthetase (the plastidic coupling factor CF_1), thereby acting as a so-called "energy-transfer" inhibitor. Using a microalga as a model system, we have been able to show that the immediate decrease of growth coincides with inhibition of ATP formation. At later stages of nitrofen treatment, other effects come into play, exerted by members of group (4), which will be reviewed here in some detail.

Using microalgae, pigment bleaching is observed with compounds of group (4) during cultivation. An early symptom is the disappearance of carotenoids and chlorophylls exhibited after application of e.g. certain phenylpyridazinones such as norflurazon, or diphenylethers like oxyfluorfen. Decrease of pigments is accompanied by a rapid loss of electron-transport activity and photosynthesis (11,12). No light-induced oxygen uptake is measurable. We have investigated some of these effects and will present data demonstrating that bleaching activity is caused by different and complex mechanisms, some of these are found also with bipyridylium herbicides. At the moment, we distinguish between two types of bleaching, either caused by inhibition of carotene biosynthesis or (peroxidative) destruction of plastidic membrane compounds.

It should be noted that early herbicidal symptoms exhibited by higher plants may be different from aquatic microalgae. Membrane destruction caused by many diphenylethers in higher plants apparently leads

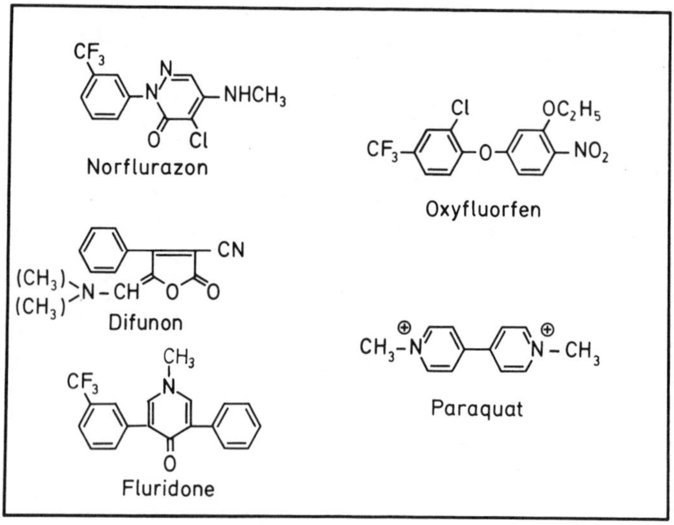

Figure 2. Chemical formulas of relevant herbicides with bleaching activity and chemical names of herbicides.

Norflurazon (SAN 9789), 4-chloro-5-methylamino-2-(3-trifluoromethylphenyl)-pyridazin-3(2H)one; SAN 6706 is the 5-dimethylamino analog (Sandoz AG).
BAS 100822, 4-chloro-5-methoxy-2-(3-tetrafluoroethoxyphenyl)-pyridazin-3(2H)-one; BAS 13338 (SAN 9785), 4-chloro-5-dimethylamino-2-phenyl-pyridazin-3(2H)-one; BAS 13761, 4-chloro-5-methoxy-2-phenyl-pyridazin-3(2H)-one; all three from BASF AG. The latter two have no bleaching activity.
Difunon (EMD-IT 5914), 5-(dimethyl-aminomethylene)-2-oxo-4-phenyl-2,5-dihydrofurane-carbonitrile-(3) (Cela-merck).
Fluridone, 1-methyl-3-phenyl-5-(3-trifluoromethylphenyl)-4(1H)-pyridinone (Eli Lilly).
Oxyfluorfen (Goal), RH 2915, 2-chloro-4-trifluoromethylphenyl-3'-ethoxy-4'-nitrophenylether; Nitrofen (TOK), 2,4-dichlorophenyl-4'-nitrophenylether;
Acifluorfen-methyl, methyl 5-[2-chloro-4-(trifluoromethyl)phenoxy]-2-nitrobenzoate (Rohm and Haas).
Paraquat (methylviologen), 1,1'-dimethyl-4,4'-bipyridylium (dichloride salt); diquat, 1,1'-ethylene-2,2'-bipyridylium.

Additional compounds with bleaching effect:

Amitrole, 3-amino-1,2,4-triazole
CGA 71884, 3,4-dimethyl-2-hydroxy-5-oxo-2,5-dihydro-N-(3-chloro-4-trifluoromethylphenyl)-pyrrolone (Ciba-Geigy)

(Legend Fig.2 continued)

Dichlormate, 3,4-dichlorobenzyl-N-methylcarbamate
DTP, 1,3-dimethyl-4-(2,4-dichlorobenzoyl)-5-hydroxy-pyrazole
Fluometuron, 3-(3-trifluoromethylphenyl)-1,1-dimethyl-urea
Haloxidine, 3,5-dichloro-2,6-difluoro-4-hydroxypyridine
Oxadiazon, 3-(2,4-dichloro-5-isopropoxyphenyl)-5-tert-butyl-1,3,4-oxadiazole-2(3H)-one
Pyrichlor, 2,3,5-trichloro-4-hydroxypyridine
R-40244, 1-(3-trifluoromethylphenyl)-3-chloro-4-chloro-methyl-2-pyrrolidone
SW 751 (Pyrazolate), 4-(2,4-dichlorobenzoyl)-5-(4-me-thylphenylsulfonyl-oxy)1,3-dimethylpyrazole

Additional herbicides mentioned in the text:

Bentazon (Basagran), 3-isopropyl-2,1,3-benzothiadia-zinone-(4)-2,2-dioxide (BASF AG)
Diuron (DCMU), 3-(3,4-dichlorophenyl)-1,1-dimethylurea; Monuron is the 4-chlorophenyl analog.

---

to rapid loss of water and/or ions. In contrast to algae, substantial bleaching cannot develop before death.

Direct effects of herbicides upon the electron transport and the membrane system are followed by secondary disturbances of plant metabolism, and many conflicting data found in the literature on primary or secondary mode of action can be resolved by carefully measuring the kinetics during herbicide treatment, using proper sublethal concentrations, different analogous derivatives, and appropriate species. Investigating modes of action of herbicides, sterile liquid cultures of unicellular microalgae have been quite advantageous. The effects are seen very early (within 24 to 48 h) and the herbicides can be applied in exact concentration to the target cell. Physiological parameters (growth, pigments, oxygen evolution) can be determined more easily than with higher plants (see [13] for details).

Interference with Carotene Biosynthesis

As shown in Table I, presence of both the phenylpyridazinone norflurazon and the diphenylether oxyfluorfen decreased the pigment content. However, a difference between these herbicides is indicated by the

Table I

Carotenoids and chlorophyll bleaching in autotrophic and heterotrophic cultures of the green microalga *Scenedesmus acutus* cultured for 24 h in the presence of 1 µM norflurazon or oxyfluorfen

Culture	Control	Norflurazon	Oxyfluorfen
Autotrophic (light)			
Chlorophyll	19.54	16.92	10.22
Carotenoids	0.053	0.023	0.015
Heterotrophic (dark)			
Chlorophyll	3.58	3.76	2.33
Carotenoids	0.011	0.005	0.008

Data are mg/ml packed cell volume. For cultivation and methods see (13).

small decrease of chlorophyll vs. carotenoids observed with norflurazon in the light, with no change of chlorophyll in the dark. In contrast, decrease of chlorophyll as well as carotenoids with oxyfluorfen present was independent of light. This dark experiment is clear evidence against a simultaneous inhibition of both carotene and chlorophyll formation by pyridazinones (14,15). Our finding is in accordance with e.g. the report of (16) that a photoflash still can induce chlorophyll synthesis in etiolated wheat leaves that were devoid of carotenes because of pretreatment with norflurazon. All data mentioned support the hypothesis that the chlorophylls are photooxidized in pyridazinone-treated plants due to the absence of carotenoids which have a protective function as natural quenchers of singlet oxygen (17); for a short overview see (18). Little ethane production is observed during the early phase of the bleaching process caused by phenylpyridazinones (Table III).

During the inhibition of carotene biosynthesis by substituted phenylpyridazinones in algae (19,20) or wheat (21) and barley (22), an accumulation of phytoene, an uncolored (triene) carotene precursor, could be observed.

Other bleaching herbicides such as the experimental phenylfuranone herbicide, difunon, (20,23) and derivatives, or fluridone (21) are thought to act on

Table II

Incorporation of radioactivity (dpm x 1000) from 1 μCi of [2-^{14}C]-DL-mevalonic acid by a Phycomyces extract in the presence of 10 μM herbicides

Herbicide	Squalene	GGPP*)	Phytoene	β-Carotene
Control	32.1	22.8	10.3	2.1
+Norflurazon	62.3	62.4	1.1	0.9
+BAS 13761	34.6	28.1	9.6	1.6
+Oxyfluorfen	35.5	26.4	11.8	2.2

*) geranylgeranyl pyrophosphate, see (26)

carotene biosynthesis very similarly to 2-phenyl pyridazinones as suggested by the accumulation of phytoene. The action of aminotriazole seems to be somewhat different. With this herbicide, ζ-carotene, γ-carotene, and lycopene are accumulated in wheat seedlings instead of phytoene (25). Consequently, an inhibition of the enzyme(s) involved in the dehydrogenation steps of carotene biosynthesis was concluded.

In Pennisetum seedlings treated with difunon in higher concentration (>10 μM), the bleaching of pigments was reported to be paralleled by a decrease of the porphobilinogenase level whereas the contents of δ-aminolevulinic synthetase and dehydratase were not lowered (24). This proposed mode of action of difunon upon chlorophyll biosynthesis could not be confirmed with the microalga Chlorella (23).

Cell-free Systems. More direct evidence can be obtained by investigating cell-free carotene biosynthesis. From Phycomyces blakesleeanus, a fungus in which β-carotene formation is inhibited by norflurazon (13), a carotenogenic enzyme system was obtained with very high ^{14}C-mevalonic acid incorporation into phytoene, and β-carotene via geranylgeranyl pyrophosphate (25). As shown in Table II, application of norflurazon inhibited radioactivity incorporation into β-carotene and phytoene. Neither BAS 13761, a non-bleaching pyridazinone, nor oxyfluorfen had an effect on ^{14}C distribution. The inhibitory site of norflurazon was found to be on the dimerization reaction of geranylgeranyl pyrophosphate. The norflurazon inhibition of phytoene synthetase from heterotrophic Phycomyces cannot explain the phytoene accumulation in green plants.

A cell-free system from photosynthetically active species has to be developed that is able to produce substantial amounts of intermediates and desaturated carotenes. Noteworthy a recent report using isolated chromoplasts from daffodil flowers, Narcissus pseudonarcissus, and isopentenyl pyrophosphate as substrate noted increased levels of phytoene and geranylgeraniol vs. control in the presence of high concentration (50 μM) of SAN 6706 (27). These cell-free assays should be investigated further with regard to possible different sensitivity against inhibitors because of the species used.

Structure/Activity Relationship. The bleaching activity of 2-phenylpyridazinones is dependent on the substituents and their position on the molecule. In a structure/activity investigation we have been able to correlate bleaching with electronic factors. Bleaching activity of 2-phenylpyridazinones is improved by substituents with increasing σ_m Hammett parameters at position 4 of the pyridazinone ring, with increasing σ_m, σ_p parameters in meta position of the phenyl moiety and with substituents in position 5 that exhibit decreasing σ_p values (28). Collectively, their primary effect is inhibition of carotene biosynthesis. According to our investigation of 2-phenylpyridazinones, analogs having a cyano or nitro group in a meta position of the phenyl ring instead of a CF_3 substituent should be potent chlorosis-inducing compounds. Indeed, bleaching comparable to SAN 6706 or norflurazon could be observed with, e.g., the 4-chloro-5-methylamino-2-(3-cyanophenyl)-pyridazinone using the green alga, Scenedesmus. This finding indicates that at least for the mode of bleaching of phenylpyridazinones, a CF_3-group is not important.

Related Effects. Other chemically different compounds are known to induce chlorosis, e.g., haloxidine and fluometuron (29), dichlormate and pyrichlor (30), SW 751 (31), the pyrrolidone derivative R 40244 (32). However, no details are known about the primary mode of action of these herbicides; dichlormate and pyrichlor most likely interfere directly with carotene formation (30). Some additional compounds with apparent bleaching activity are mentioned in (30,31).

Besides the bleaching caused by either carotene inhibition or peroxidation more targets for "bleaching herbicides" are conceivable, e.g., interference with chlorophyll formation itself. DTP, a substituted pyrazole, was reported to induce chlorosis by blocking synthesis of protochlorophyllide (33). Oxadiazon, an oxadiazolone herbicide, induces a strong bleaching with autotrophic Scenedesmus (I_{50} around 10^{-7}M) which is neither accompanied by formation of phytoene nor volatile hydrocarbons (this laboratory, unpubl. results). More research will have to clarify their mode of action.

Primary herbicidal effects are followed by secondary ones that show up before death of the plant cell. The 70-S ribosomes of wheat chloroplasts are decreased by bleaching pyridazinones in the light, but not in the dark (29). A prominent mode of action is observed upon the composition of fatty acids by, e.g., BAS 13338 (SAN 9785) (34,35), which does not substantially interfere with carotenoid biosynthesis. Good direct inhibition of photosynthetic electron transport (I_{50} = 3 x 10^{-7}M) is observed with the phenylpyridazinone BAS 100822; electron transport inhibition of other phenylpyridazinones is less than with BAS 100822 (28).

Herbicide-induced Peroxidations

Diphenylethers. Different agents can induce peroxidations in photosynthetic membranes. The strongest peroxidative herbicides known are the p-nitrodiphenyl ethers. Although compounds from this group of herbicides can act multifunctionally as electron-transport inhibitors (2,36) and energy-transfer inhibitors (10), some of them, like oxyfluorfen, exert their dominant phytotoxic action by damaging membrane components. Also nitrofen has a strong peroxidative effect in addition to energy-transfer inhibition.

Markers to follow membrane degradation are chlorophyll and carotenoid content of the culture (11) or volatile short-chain hydrocarbon formation during peroxidation of fatty acids (12,37). The kinetics of all three parameters in oxyfluorfen-treated algal cells is demonstrated in Figure 3. Ethane formation was also observed with isolated spinach chloroplasts (37). As was further demonstrated (12; Figure 3), photosynthetic electron transport decayed rapidly under the influence of oxyfluorfen although its direct inhibition is not possible with the concentrations applied.

As demonstrated by Table III, norflurazon-treated Scenedesmus evolved only traces of ethane after 4 hrs of incubation, while presence of the diphenylethers oxyfluorfen and acifluorfen-methyl (1 µM each) produced substantial amounts reaching a maximum after 18 to 20 hrs. At this time, carotenoids had decreased by

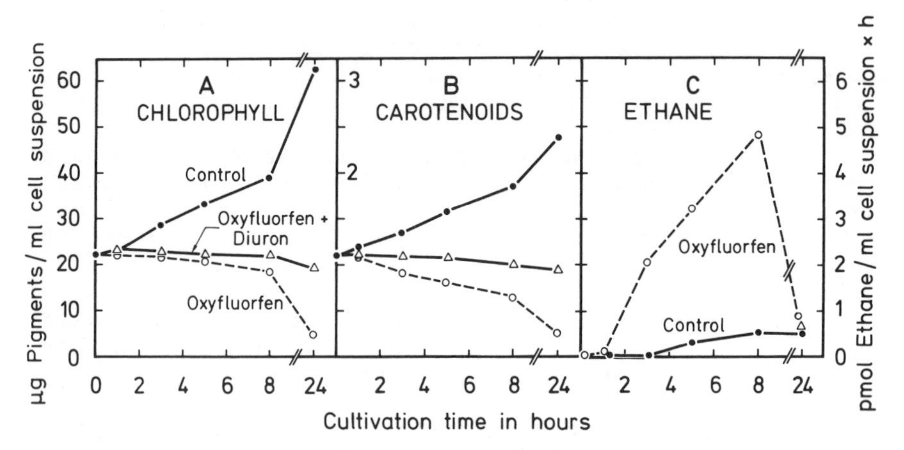

Figure 3. A,B: Changes of chlorophyll and carotenoid content of the microalga Scenedesmus acutus *under the influence of oxyfluorfen (1 µM). Curve △ indicates alleviation of oxyfluorfen activity when 1 µM DCMU (diuron) is given simultaneously. C: Light-induced ethane formation during treatment of the cells with 1 µM oxyfluorfen over 24 h (*12*).*

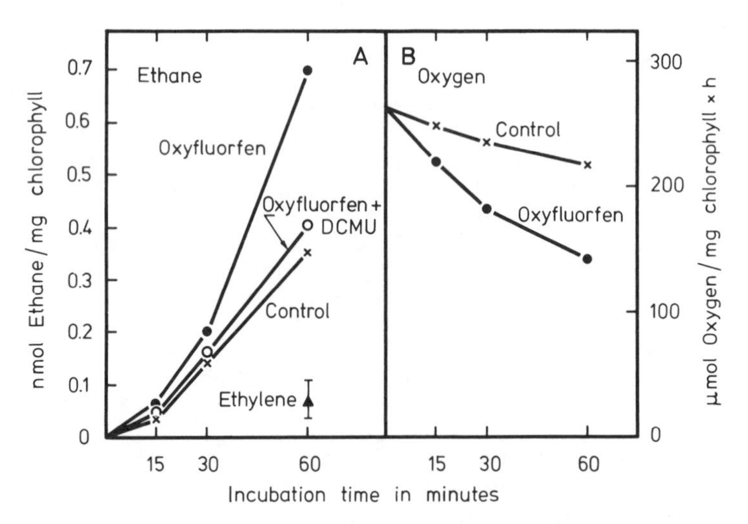

*Figure 4. A: Light-induced formation of ethane and ethylene over 60 min with isolated spinach chloroplasts in the presence of 1 µM oxyfluorfen and 1 µM oxyfluorfen combined with 1 µM DCMU (diuron). B: Decrease of the Hill reaction (oxygen evolution: system $H_2O \rightarrow$ potassium ferricyanide) during incubation of isolated chloroplasts in the light with 1 µM oxyfluorfen present (*12, 37*).*

Table III

Ethane formation in diphenylether- and norflurazon-treated Scenedesmus

Herbicide (1 µM)	Time of treatment		
	0	4 h	18 h
(1) Oxyfluorfen	0	2.7	4.9
(2) Oxyfluorfen (red light)	0	1.1	5.6
(3) Oxyfluorfen + diuron, 1 µM	0	0.1	0.8
(4) Acifluorfen-methyl	0	3.0	17.8
(5) Norflurazon	0	0.2	1.9

Data are nmol/ml packed cell volume. Gas measurement was performed after the herbicide treatment over a 1-h illumination period with white or red light (>610 nm, line 2) of 100 W/m^2 according to (64). Dark controls were below 5% of the data measured in the light.

about the same degree (Table I), while ethane formation of the norflurazon-treated cells was still low, although it had somewhat increased due to peroxidations and/or chlorophyll-sensitized photooxidations developing in the course of inhibited carotene biosynthesis. As demonstrated further with intact cells (Figure 3; Table III, line 3) and isolated chloroplasts as well (Figure 4), ethane evolution induced by oxyfluorfen decreased when electron transport was blocked by simultaneous addition of diuron (12). Ethane evolution was lowered likewise with oxyfluorfen given in the dark. Noteworthy, red light beyond 610 nm gave results comparable to those obtained with white light (Table III, lines 1,2) indicative of carotenoids being not directly involved in diphenylether activation (38).

The requirement of light for the activation of oxyfluorfen and nitrofen has been known for years. However, the involvement of photosynthetic electron transport was doubted (39,40). At the moment, our hypothesis claims p-nitrophenylethers to be reduced to a (nitroanion?) radical which subsequently initiates peroxidation of (galactolipid) polyunsaturated fatty acids and other membrane components. Apparently, also respiratory reactions are able to activate oxyfluorfen, since our heterotrophically growing cultures exhibit a small bleaching in the dark (Table I). Possibly these

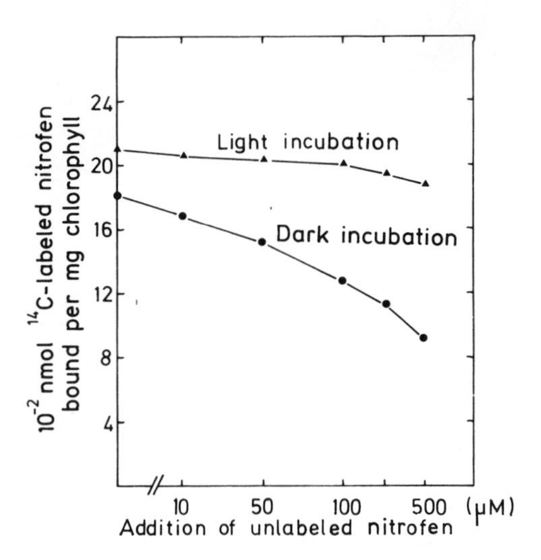

Figure 5. Addition of nitrofen to isolated spinach chloroplasts in the light and in the dark. Thylakoids equivalent to 100 μg of chlorophyll were incubated at pH 7.4 in the presence of 50 μM ^{14}C-labeled nitrofen for 15 min. Aliquots of this preloaded material were then treated with increasing concentrations of cold nitrofen as indicated (37)*.*

radicals also bind to the thylakoid membrane. Illumination of isolated spinach chloroplasts in the presence of ^{14}C-labeled nitrofen led to a strong attachment of this diphenylether (Figure 5); no replacement by unlabeled nitrofen was possible (37).

Our experiments were performed with microalgae under optimum conditions of photosynthesis such as supply of light, CO_2, minerals, etc. It is conceivable that suboptimum conditions may allow for oxygen activation at the pigments leading to subsequent activation of (bleaching) diphenylethers. Carotenes may then be functional for the action of diphenylethers as was recently confirmed with higher plants treated with acifluorfen-methyl (41). Using *Scenedesmus*, photosynthetic electron transport was not necessary *after* a 2-hr oxyfluorfen treatment (12) when photosynthesis started to decay anyway. Then surplus of light energy possibly may be diverted directly via piments as indicated above.

Bipyridyls. Biypridylium salts like paraquat also cause destruction of chloroplast organization (42). An early symptom is the light-induced oxygen uptake and the strong inhibition of the photosystem-II region with photosystem I not being affected. Concurrently, malondialdehyde is formed. A decrease of chlorophyll is not observed to occur concurrently (9). Ethane can be formed under the influence of paraquat (43,44). In two recent publications it was shown by fatty acid determinations that unsaturated fatty acids are preferentially degraded under the influence of paraquat (45, 46). The primary mode of action is its one-electron reduction and the subsequent reduction of the radical by molecular oxygen (7). The resulting superoxide anion is assumed to give rise to other active oxygen species (8). Apparently, light-induced paraquat reduction and subsequent peroxidation is also achieved without electron transport, e.g., in subchloroplast particles where endogenous substrates or chlorophyll itself may be the electron donors (47). Similar peroxidative processes are induced by diquat (48).

Mechanism. Degradation of α-linolenic acid (α-lin) as proposed by (49,50,51) is demonstrated in Figure 6. The initial step is a hydrogen abstraction from an α-linolenic molecule by a radical species that was formed as a result of herbicidal action. In the following radical-chain reaction the ω-3 alkyl peroxide is formed via the peroxy radical. Subsequently, this peroxide is decomposed in a Fenton-type reaction to the ω-3 alkoxy radical in the presence of transition metals that can undergo one-electron transfer reaction, e.g., Cu(I/II), Fe(II/III), Ti(III/IV), or Ce(III/IV). The ω-3 alkoxy radical can split by β-scission to an unsaturated aldehyde and the ethyl radical. The latter is either oxidized to ethylene or reduced to ethane.

Quantitative estimation indicates that only about 1% of peroxidized α-lin is decomposed to ethane or ethylene (52). However, the high sensitivity of gas-chromatography for volatile hydrocarbons and the advantage that this determination can be done with intact organisms make short-chain hydrocarbons excellent markers to trace herbicide-induced peroxidation reactions.

In addition to ethane/ethylene, other volatile hydrocarbons can be formed in substantial amounts. The xanthophycean alga *Bumilleriopsis filiformis* produces mainly propane after treatment with paraquat (44) or oxyfluorfen (12), whereas lipid peroxidation in animals results in pentane formation (53).

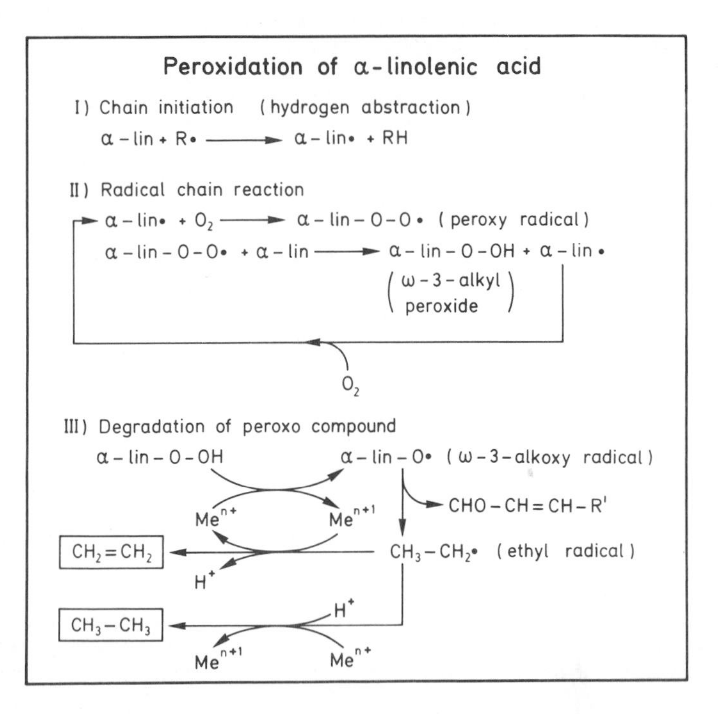

*Figure 6. Peroxidation of a ω-3-polyunsaturated fatty acid. A radical R· generated, e.g., by the photosynthetic electron-transport system, can abstract a hydrogen from α-linolenic acid (α-lin). After conjugation of two double bonds, this leads to linolenic acid radical (α-lin·), which reacts with molecular oxygen giving peroxy radical. This again reacts with another linolenic acid molecule thereby starting the chain reaction (part II). The resulting ω-3-alkyl peroxide is reduced to ω-3-alkoxy radical through a Fenton-type reaction (*see *part III) mediated by metal ions like Fe(II/III) or Cu(I/II). By β-scission this radical splits into the corresponding aldehyde and ethyl radical (R′ denotes the fatty acid moiety carrying the carboxyl group). With copper ions present, the ethyl radical reacts with Cu(II) to give ethylene and with Cu(I) to yield ethane. With Cu(II) in excess, most of the hydrocarbon produced is ethylene (*51*). We assume that most of the Cu(II) is reduced by photosynthetic electron transport.*

Origin of Hydrocarbon Gases. In order to find out which fatty acids are the sources of volatile hydrocarbons we used thylakoid preparations from several blue-green algae that contain different fatty acids. The peroxidative reactions were mediated and sustained by Cu(II) ions in the light. Table IV shows that fatty acids with at least 2 double bonds are necessary for hydrocarbon formation. As *Anacystis* lacks those fatty acids, no peroxidative volatile hydrocarbons were produced. *Spirulina* exclusively contains ω-6 fatty acids as endogenous polyunsaturated fatty acids and evolved C_5 hydrocarbons only. The third species, *Anabaena*, whose thylakoids contain ω-3 and ω-6 polyunsaturated fatty acids formed C_2 and C_5 hydrocarbons simultaneously. We conclude that ω-3 polyunsaturated fatty acids are the source of ethane and ethylene and that the ω-6 polyunsaturated fatty acids are the source of pentane and pentene in herbicide-induced peroxidation reactions. Furthermore, we obtained evidence that the propane measured with *Bumilleriopsis* after an 18-h treatment with either 10 µM oxyfluorfen or 50 µM Cu(II) originates from a ω-4 polyunsaturated fatty acid. We have recently isolated and identified this acid as 16:3ω4 (Sandmann, Lambert, Böger; in preparation).

This biological plant system confirms and extends the report of Dumelin and Tappel (52) who were the first to prove the origin of ethane and pentane by a metal- or hematin-catalyzed chemical reaction. Furthermore, addition of appropriate fatty acids to isolated thylakoids of *Anacystis* according to Table IV, after short-term illumination yielded hydrocarbon evolution that confirmed the rule explained above (54). Using the latter system we could also demonstrate that hydrocarbons do not originate from carotenoids. Experiments with isolated fatty acids, however, should be done in short experimental times, otherwise unphysiological peroxidations of fatty acids may occur (comp. 55).

Additional Assays on Herbicidal Mode of Bleaching

Direct evidence for degradation of lipids can be obtained by prelabeling the thylakoid sulfolipid of, e.g., microalgae with ^{35}S-sulfate. When oxyfluorfen or norflurazon were applied to cause a 90% decrease of α- and β-carotene over 2 culture days, more than 90% of the radioactivity of the sulfolipid disappeared in the oxyfluorfen sample. Using norflurazon, only 20% of the sulfolipid was degraded during this period (56).

Table IV

Concentration of polyunsaturated fatty acids in the thylakoids of blue-green algae and light-induced volatile hydrocarbon production by spheroplasts in the presence of 50 µM Cu(II)

	Anacystis nidulans	Anabaena variabilis	Spirulina platensis
Fatty acids, µg/µg chlorophyll			
16:2 ω6	0	0	0
16:3 ω3	0	0.32	0
18:2 ω6	0	14.9	9.2
18:3 ω3	0	19.9	0
18:3 ω6	0	0	12.8
20:4 ω6	0	0	0
Hydrocarbons produced, pmol/mg chlorophyll x h			
Ethane	0	109.8	0
Ethylene	0	113.2	0
Pentane	0	131.3	92.5
Pentene	0	45.5	69.8

Spheroplasts prepared according to (65) were shocked and illuminated for 2 hours, according to (51,54).

A rapid screening method to distinguish between an inhibitory mode of action on carotenoid biosynthesis and peroxidative bleaching is the measurement of fluorescence induction (57). In the presence of both norflurazon and oxyfluorfen, the variable fluorescence signal of *Scenedesmus* was abolished after an 8-h treatment. Based on chlorophyll, the maximum fluorescence-signal height of norflurazon-treated cells rose about twice as high as the control, whereas that of the oxyfluorfen-treated cells decreased to about 10%. Furthermore, only with the pyridazinone-treated cells was the constant fluorescence increased, as was seen by a "spike"-type signal absent in the oxyfluorfen sample. Undoubtedly, these fluorescence-signal changes are caused by the herbicide-induced decay of electron-transport activity. According to the signal differences, this decay is apparently produced in a different manner due to disturbed energy transfer, pigment destruction or degradation of redox carriers.

Peroxidations as Secondary Herbicidal Effects

As pointed out previously for chlorophyll destruction and ethane evolution with algae in the presence of norflurazon (Tables I,III), peroxidative reactions may show up as a consequence of a primary inhibition of carotene synthesis, which itself does not induce peroxidation-initiating radicals.

In addition to the bleaching herbicides, electron transport inhibitors binding at the B protein (Figure 1) may initiate peroxidation after long-term treatment and under high light intensity. As a consequence of inhibited electron transport, the excited chlorophylls transfer their energy to ground state oxygen yielding highly reactive singlet oxygen (43,58). Singlet oxygen is quenched by carotenes. Apparently, carotenes also are broken down in this reaction preceding the destruction of chlorophyll (11,59,60). When the carotene level is decreased, photodestruction of chlorophyll as well as lipid peroxidation are started as observed with peroxidative diphenylethers. Such pigment degradations are known for monuron (59), diuron (58,60), and the experimental phenyl-hydroxy-pyrrolone herbicide CGA 71884 (unpublished results). This compound is an electron-transport inhibitor with $I_{50} = \cdot 7 \times 10^{-7}$ M (61).

These secondary bleaching phenomena generally require quite unphysiological conditions such as substantially inhibited electron transport and/or a decreased carotene level in the thylakoids (induced, e.g., by phenylpyridazinones) together with a high quantum flux reaching the pigments. By carefully choosing the cultivation conditions, e.g., allowing for some growth with sublethal concentrations of s-triazines or triazinone herbicides present, even increased levels of chlorophyll may be observed ("greening effect", see [62] for higher plants, [63] for algae).

Acknowledgment

This study was supported by the Deutsche Forschungsgemeinschaft. The experiments were made possible through a generous supply of pure herbicide samples which were gifts from BASF AG, Limburgerhof and Ludwigshafen, Celamerck GmbH, Ingelheim, both Germany, Ciba-Geigy AG, Sandoz AG, both Basel, Switzerland, and Rohm and Haas, Spring House, Pennsylvania, USA (see legend of Figure 2).

Literature Cited

1. Moreland, D.E. Annu. Rev. Plant Physiol. 1980, 31, 597-638.
2. Böger, P. Plant Res. Development (Tübingen) 1978, 8, 79-101.
3. Pfister, K.; Arntzen, C.J. Z. Naturforsch. 1979, 34c, 996-1009.
4. Trebst, A.; Draber, W. in "Advances in Pesticide Science" Part 2; Ed. H. Geissbühler, Pergamon Press, Oxford-New York, 1979; pp. 223-234.
5. Böger, P. Z. Pflanzenkr. Pflanzenpathol. Pflanzenschutz (Plant Disease and Protection) 1981, spec. issue no. IX, 153-162.
6. Black, C.C Biochim. Biophys. Acta 1966, 120, 332-340.
7. Farrington, J.A.; Ebert, M.; Land, E.J.; Fletcher, K. Biochim. Biophys. Acta 1973, 314, 373-381.
8. Dodge, A.D. in "Herbicides and Fungicides"; Ed. N.R. McFarlane, Chemical Society, Special Publication 29, London, 1977; pp. 7-21.
9. Böger, P.; Kunert, K.J. Z. Naturforsch. 1978, 33c, 688-694.
10. Lambert, R.; Kunert, K.J.; Böger, P. Pesticide Biochem. Physiol. 1979, 11, 267-274.
11. Kunert, K.J.; Böger, P. Z. Naturforsch. 1979, 34c, 1047-1051.
12. Kunert, K.J.; Böger, P. Weed Sci. 1981, 29, 169-173.
13. Sandmann, G.; Kunert, K.J.; Böger, P. Z. Naturforsch. 1979, 34c, 1044-1046.
14. Lichtenthaler, H.K.; Kleudgen, H.K. Z. Naturforsch. 1979, 32c, 236-240.
15. Kleudgen, H.K. Pest. Biochem. Physiol. 1979, 12, 231-238.
16. Axelsson, L.; Klockare, B.; Ryberg, H.; Sandelius, A.S. Proc. 5th Int. Congr. Photosynth. Halkidiki, Greece, 1981, in press.
17. Foote, C.S. in "Free Radicals in Biology" Vol. II; Ed. W.A. Prior, Academic Press, New York, 1976; pp. 85-133.
18. Eder, F.A. Z. Naturforsch. 1979, 34c, 1052-1054.
19. Vaisberg, A.J.; Schiff, J.A. Plant Physiol. 1976, 57, 260-269.

20. Urbach, D.; Suchanka, M.; Urbach, W. Z. Naturforsch. 1976, 31c, 652-655.
21. Bartels, P.G.; Watson, C.W. Weed Sci. 1978, 26, 198-203.
22. Ridley, S.M.; Ridley, J. Plant Physiol. 1979, 63, 392-398.
23. Kunert, K.J.; Böger, P. Weed Sci. 1978, 26, 292-296.
24. Hampp, R.; Sankhla, N.; Huber, W. Physiol. Plant. 1975, 33, 53-57.
25. Rüdiger, W.; Benz, J.; Lempert, U.; Schoch, S.; Steffens, D. Z. Pflanzenphysiol. 1976, 80, 131-143.
26. Sandmann, G.; Bramley, P.M.; Böger, P. Pest. Biochem. Physiol. 1980, 14, 185-191.
27. Beyer, P.; Kreuz, K.; Kleinig, H. Planta (Berl.) 1980, 150, 435-438.
28. Sandmann, G.; Kunert, K.J.; Böger, P. Pest. Biochem. Physiol. 1981, 15, in press.
29. Feierabend, J.; Schubert, B. Plant Physiol. 1978, 61, 1017-1022.
30. Burns, E.F.; Buchanan, G.A.; Carter, M.G. Plant Physiol. 1971, 47, 144-148.
31. Draber, W.; Fedtke, C. in "Advances in Pesticide Science"; Ed. H. Geissbühler, Pergamon Press, Oxford-New York, 1979, pp. 475-486.
32. Devlin, R.M.; Kisiel, M.J.; Kostusiak, A.S. Weed Res. 1979, 19, 59-61.
33. Kawakubo, K.; Shindo, J. Plant Physiol. 1979, 64, 774-779.
34. St.John, J.B.; Christiansen, M.N. Plant Physiol. 1976, 57, 38-40.
35. Willemot, C. Plant Physiol. 1977, 60, 1-4.
36. Bugg, M.W.; Whitmarsh, J.; Rieck, C.E.; Cohen,W.S. Plant Physiol. 1980, 65, 47-50.
37. Lambert, R.; Böger, P. Z. Pflanzenkr. Pflanzenschutz (Plant Disease and Protection) 1981, spec. issue no. IX, 147-152.
38. Lambert, R.; Sandmann, G.; Böger, P. Weed Sci. 1982, submitted.
39. Matsunaka, S. Residue Rev.1969, 25, 45-58.
40. Fadayomi, O.; Warren, G.F. Weed Sci. 1976, 24, 598-600.
41. Orr, G.K.; Hess, F.D. Abstracts 1981 Meeting WSSA, No. 248.
42. Harris, N.; Dodge, A.D. Planta (Berl.) 1972, 104, 210-219.
43. Elstner, E.F. Ber. Dtsch. Bot. Ges. 1978, 91, 569-577.

44. Boehler-Kohler, B.A.; Läpple, G.; Hellmann, V.; Böger, P. Pestic. Sci. 1981, in press.
45. Youngman, R.J.; Dodge, A.D. Z. Naturforsch. 1979, 34c, 1032-1035.
46. Boehler-Kohler, B.A.; Schopf, M.; Böger, P. Abstract, 5th Int. Congr. Photosynth. Halkidiki, Greece 1980.
47. Elstner, E.F.; Lengfelder, E.; Kwiatkowski, G. Z. Naturforsch. 1979, 35c, 303-307.
48. van Rensen, J.J.S. Physiol. Plant. 1975, 33, 42-46.
49. Halliwell, B. in "Strategies of Microbial Life in Extreme Environments", Life Sciences Research Report 13; Ed. M. Shilo, Verlag Chemie, Weinheim, 1979, pp. 195-221.
50. Elstner, E.F.; Pils, I. Z. Naturforsch. 1979, 34c, 1040-1043.
51. Sandmann, G.; Böger, P. Plant Physiol. 1980, 66, 797-800.
52. Dumelin, E.E.; Tappel, A.L. Lipids 1977, 12, 894-900.
53. Dillard, C.J.; Dumelin, E.E.; Tappel, A.L. Lipids 1977, 12, 109-144.
54. Sandmann, G.; Böger, P. Lipids 1982, in press.
55. Schobert, B.; Elstner, E.F. Plant Physiol. 1980, 66, 215-219.
56. Sandmann, G.; Böger, P. Plant Sci. Lett. 1981, in press.
57. Böhme, H.; Kunert, K.J.; Böger, P. Weed Sci. 1981, in press.
58. Takahama, U.; Nishimura, M. Plant Cell Physiol. 1975, 16, 737-748.
59. Pallett, K.E.; Dodge, A.D. Z. Naturforsch. 1979, 34c, 1058-1061.
60. Ridley, S.M. Plant Physiol. 1977, 59, 724-732.
61. Brugnoni, G.P.; Moser, P.; Trebst, A. Z. Naturforsch. 1979, 34c, 1028-1031.
62. Fedtke, C. Weed Sci. 1979, 27, 192-195.
63. Böger, P.; Schlue, U. Weed Res. 1976, 16, 149-154.
64. Sandmann, G.; Böger, P. Z. Pflanzenphysiol. 1980, 98, 53-59.
65. Spiller, H.; Böger, P. Methods Enzymol. 1980, 69, 105-115.

RECEIVED September 15, 1981.

8

Proposed Site(s) of Action of New Diphenyl Ether Herbicides

GREGORY L. ORR
Colorado State University, Department of Botany and Plant Pathology, Fort Collins, CO 80523

F. DANA HESS
Purdue University, Department of Botany and Plant Pathology, West Lafayette, IN 47907

The new diphenylether (DPE) herbicides, e.g., acifluorfen-methyl {methyl 5-[2-chloro-4-(trifluoromethyl)phenoxy]-2-nitrobenzoate} are proposed to be activated in light by carotenoids and then initiate radical chain reactions with membrane fatty acids. This hypothesis is based on the following: (a) DPE's are active in green and etiolated tissue; (b) damage does not occur after inhibition of carotenoid biosynthesis by fluridone {1-methyl-3-phenyl-5-[3-(trifluoromethyl)phenyl]-4(1H)-pyridinone}; (c) incurred damage requires light and oxygen; (d) injury is expressed as a general increase in membrane permeability 10 to 15 min following light-activation; (e) membrane disruption can be verified by electron microscopy; (f) ultrastructural analysis revealed early injury to the chloroplast envelope; (g) ethane, ethylene, and thiobarbituric acid-reacting materials can be detected after treatment; (h) pretreatment with α-tocopherol can protect against DPE injury.

To determine the physiological and biochemical mechanism of action of a given group of herbicides, the primary site of action must be identified. This is often difficult when the physiological effect requires long periods of time to detect or is induced only by high concentrations. There are now several DPE's available that exhibit a rapid and relatively high degree of herbicidal activity. Thus, following the development of bioassays capable of detecting subtle quantitative differences between biological responses induced by these herbicides, it became possible to study the primary herbicidal mechanism.

In the following discussion, much information concerning the biochemical mechanism of action of DPE's will be from research on acifluorfen-methyl (AFM). However, we will also review recent developments from studies of other compounds

0097-6156/82/0181-0131$05.25/0

in this class, e.g., oxyfluorfen [2-chloro-1-(3-ethoxy-4-nitrophenoxy)-4-(trifluoromethyl)benzene], bifenox [methyl 5-(2,4-dichlorophenoxy)-2-nitrobenzoate]. Finally, we shall examine, from a theoretical viewpoint, proposals for future research and speculate on possible outcomes and interpretations.

Structure-activity Relationships

The rate of efflux of $^{86}Rb^+$ from excised and preloaded cucumber (*Cucumis sativus* L.) cotyledons (Figure 1), was used to study structure-activity relationships for several DPE herbicides (1). It was apparent an R-group ortho to the nitro group on the *p*-nitrophenyl moiety was required for short-term (<8 h) expression of herbicidal activity in the cucumber cotyledon bioassay system (1). However, the nature of the R-group was also important. There was a decrease in activity when the R-group at this position was changed from a methyl ester (AFM) to an ethyl ester (acifluorfen-ethyl {ethyl 5-[2-chloro-4-(trifluoromethyl) phenoxy]-2-nitrobenzoate}) to an ethyl ether (oxyfluorfen). Compounds containing R-groups with an electrical charge (e.g., the salt of acifluorfen) were inactive in this assay. The inactivity may be due to inadequate penetration. If DPE's initiate radical chain reactions within cell membranes, an R-group with an electrical charge may prevent the molecule from partitioning into the phospholipid bilayer deeply enough to interact with the polyunsaturated fatty acid (PUFA) moieties. The activity of bifenox could be significantly increased merely by substituting the chlorine atom at the 4 position of the 2,4-dichlorophenoxy moiety with a trifluoromethyl group (AFM).

Use of the cucumber cotyledon bioassay is limited because of probable differences in herbicide penetration and translocation. If all DPE herbicides affect the same site(s), then discrepancies in activity could also be related to the differential ability of cucumber cotyledons to detoxify these compounds. Further structure-activity studies must utilize *in vitro* assays to measure intrinsic activity.

The differences in relative activity of these compounds may be related to effects various constituents have on the ability to form highly reactive free radicals. A study which could help elucidate the actual mechanism of action of these compounds would be to examine (theoretically or experimentally) their ability to form radicals. If they express their herbicidal activity via this mechanism, then the energy and reactivity of free radical formation should show a positive correlation with the relative activity differences found in an intrinsic bioassay. For this purpose, the midpoint redox potential of the one-electron transfer couple $DPE\text{-}NO_2 \rightarrow DPE\text{-}NO_2^-$ needs to be determined. This could be accomplished using such techniques as polarography or cyclic voltometry. It may also be feasible to study these reactions in darkness through the use of artificial electron

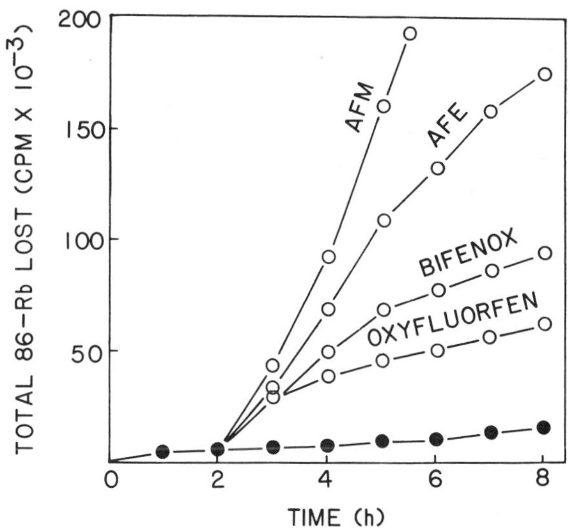

Figure 1. Efflux of $^{86}Rb^{+}$ from cucumber cotyledons in the presence of 1 μM DPE herbicides (1). At time zero, cotyledons were exposed to 600 μE/m ^{2} s (PAR) light and herbicide. Closed circles are efflux from untreated cotyledons. Abbreviations: AFM, acifluorfen-methyl; AFE, acifluorfen-ethyl.

donors. By stabilizing the radicals with various spin adducts (i.e., spin trapping agents), it should be possible to observe and assign the resulting signal using electron spin resonance (ESR) spectroscopy (2).

Characterization

Many DPE herbicides produce injury upon contact with plant foliage. Injury symptoms become evident as a "water-soaked" appearance. This is generally thought to result from a leakage of cellular constituents into the intercellular air spaces of the leaf.

By measuring the rate of $^{86}Rb^+$ efflux from excised and preloaded cucumber cotyledons, AFM injury was detected at concentrations as low as 10 nM (1). Also, because AFM has an absolute requirement of light for expression of herbicidal activity, plant tissues can be pretreated in darkness without injury. Then, following light-activation, damage can be detected in relatively short time periods (10 to 15 min) (1). We believe these observations are indicative of its primary biochemical mechanism of action.

Light Requirement and Membrane Disruption. As mentioned above, DPE's have an absolute requirement of light for activity (1, 3-12). The light-activated form of the AFM molecule apparently has a relatively short half-life, because further injury can be prevented by returning the tissue to darkness (1). By decreasing light quantity, the effect of AFM is delayed, although magnitudes of the responses at different intensities are nearly equal (1). The same observations were made with oxyfluorfen (10).

DPE treatment appears to disrupt cell membranes. Following light-activation of AFM, injury can be detected as: (a) a general increase in electrolytic conductivity of the external bathing solution of treated tissues; (b) efflux of inorganic ions such as $^{86}Rb^+$, $^{36}Cl^-$ (Figure 2A), and $^{45}Ca^{2+}$ (Figure 2B); (c) efflux of a neutral organic molecule, 3-O-methyl-[^{14}C]glucose (Figure 2C); and (d) efflux of a charged organic molecule, [^{14}C]methylamine$^+$ (Figure 2D).

Potential Resistance Mechanism(s). In studies designed to elucidate the mechanism of action of AFM, cucumber seedlings were grown in the dark for 6 or 7 days. Cotyledons were excised, greened, and loaded with $^{86}Rb^+$ in low light (75 $\mu E\ m^{-2}\ s^{-1}$) for 24 h. The cotyledons were subsequently exposed to herbicide in high light (600 $\mu E\ m^{-2}\ s^{-1}$) and found to be injured rapidly by micromolar concentrations of AFM (1). However, excised cotyledons allowed to green in low light for 48 h, before high light and herbicide treatment, were not injured. Resistance may be attributed to decreased penetration resulting from an

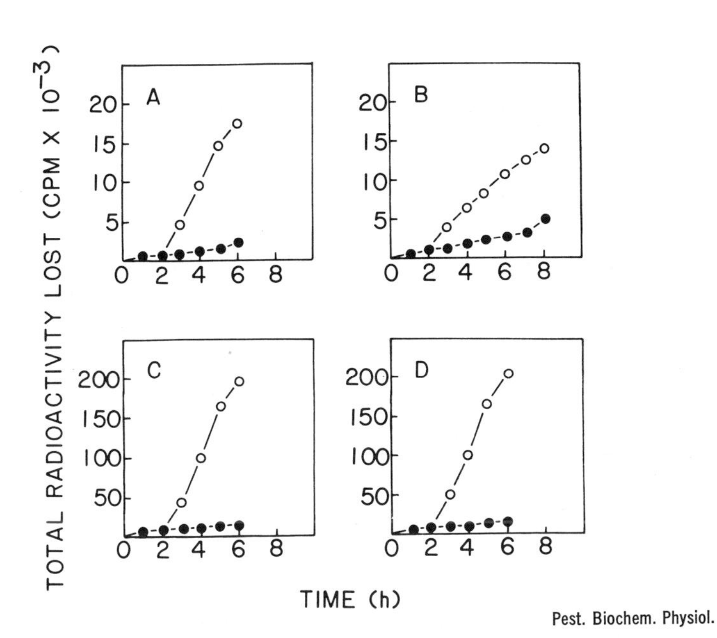

Pest. Biochem. Physiol.

Figure 2. Effect of 1 μM AFM on efflux of various radiolabeled substances from cucumber cotyledons (1). At time zero, cotyledons were exposed to 600 μE/m² s (PAR) light and herbicide. A, Efflux of $^{36}Cl^-$. B, Efflux of $^{45}Ca^{2+}$. C, Efflux of 3-O-methyl-[^{14}C]glucose. D, Efflux of [^{14}C]methylamine$^+$. Closed circles are efflux from untreated cotyledons.

increase in cuticle thickness [as was found (13) for cabbage (*Brassica oleraceae* var. capitata L.) treated with nitrofen (2,4-dichlorophenyl *p*-nitrophenyl ether)]. Increased levels of free carbohydrates, derived through the glyoxylate cycle from fats stored in lipid bodies, may account for resistance by forming various herbicide-glycoside complexes. Mature green tissues may contain detoxification mechanisms not present in immature tissues (14).

The final resistance mechanism to be discussed is related more directly to the proposed mechanism of action of the newer DPE's. Tissue allowed to green for 48 h before herbicide treatment may have increased levels of *in vivo* free radical scavengers (e.g., tocopherols) that are able to quench the light-activated form of the DPE molecule or subsequent reaction products (15). This resistance mechanism might be overcome by destroying the α-tocopherol (α-T) present with picrylhydrazyl (16), a nitrogen radical known to specifically react with α-T in lipid mixtures.

DPE's have shown herbicidal activity in a *Chlamydomonas* bioassay (17). *Chlamydomonas eugametos* cultures were grown under a 12:12 h light:dark cycle for 4 days and then treated with 1 μM AFM. A complete inhibition of cell population increase was observed 24 h later. However, there was no significant increase in efflux of $^{86}Rb^{+}$ even after 6 to 8 h of exposure. Because tocopherols are known to be synthesized by plants after exposure to light (15), this short term protection against AFM may be caused by increased quantities of radical quenching compounds. This protective mechanism is essential for plant life in the presence of light and oxygen, which create the potential for tissue damage by free radicals. Therefore, treatment of the alga with picrylhydrazyl may decrease the time required to observe herbicidal injury symptoms.

Once α-T quenches a radical and becomes an aroyl radical, there is evidence it is reconverted (repaired) to α-T *in vivo* by thiols (18). Therefore, it may be possible to decrease the induction period for membrane damage introduced by *in vivo* α-T through destruction of glutathione (a thiol thought to be involved in the repair mechanism). This could be accomplished by pretreating cucumber cotyledons greened for 48 h with a compound, such as diethyl maleate, which has been shown to destroy glutathione (19). Therefore, by decreasing the levels of glutathione, tissue would lose the ability to prevent the formation of lipid peroxides or to detoxify lipid peroxides once formed. Any compound increasing the cellular content of thiols (e.g., glutathione) would be a candidate for an antidote of AFM. One possible protectant against DPE injury to corn (*Zea mays* L.) is R-25788 (*N*,*N*-diallyl-2,2-dichloroacetamide). Tocopherols are unsuited for field use because they are degraded in light and air (20).

Possible DPE Effects on Phenolics. Phenolic levels were significantly increased in AFM treated tissue (unpublished results). This may be an indication of a wound response. However, it is also possible these herbicides may regulate the activity of one of the light-activated enzymes involved in phenolic acid synthesis. The resultant increase in high levels of free radical intermediates known to occur in these pathways, could be the ultimate cause of cellular destruction. Treatment of cucumber with fluorodifen [*p*-nitrophenyl (α,α,α-trifluoro-2-nitro-*p*-tolyl)urea] has been shown to increase phenylalanine ammonia lyase (PAL) activity *in vivo* (21).

Light-activating Mechanism(s)

The activation of DPE's by light appears to require neither chlorophyll nor photosynthetic electron transport (3). AFM will induce herbicidal injury in green and etiolated cucumber cotyledons in the presence of DCMU [3-(3,4-dichlorophenyl)-1, 1-dimethylurea] and DBMIB (2,5-dibromo-3-methyl-6-isopropyl-*p*-benzoquinone), *in vivo* inhibitors of non-cyclic and cyclic photosynthetic electron transport, respectively. Recently, Bugg *et al*. (22) obtained evidence that indicated the site of photosynthetic electron transport inhibition by nitrofluorfen [2-chloro-1-(4-nitrophenoxy)-4-(trifluoromethyl)benzene], was associated with the plastoquinone-Cyt f region between PS I and PS II. This is in agreement with previous research conducted on the mechanism of action of DPE's (13, 23-27). However, in view of the above data from cucumber, these results are probably not indicative of the primary herbicidal site of action.

After 2 to 3 h, etiolated tissues exposed to AFM and high light (600 $\mu E\ m^{-2}\ s^{-1}$) show typical injury symptoms and significant increases in the rate of $^{86}Rb^{+}$ efflux (3). The chorophyll content in the etiolated control tissues, even after 4 h in light, was less than 1% of the green tissues. Although this analysis indicated no quantitative loss of pigments, tissue bleaching was apparent (i.e., the oxidation and subsequent loss of pigments was visually evident on the periphery of the leaf). Etiolated cucumber cotyledons examined with the electron microscope revealed only etioplasts with large prolamellar bodies.

Potential Involvement of Carotenoids. The pigment involved in the light-activating mechanism of the DPE molecule may be a carotenoid (5, 6, 7). The absorption spectrum of a crude pigment extract taken from etiolated cucumber cotyledons very closely matches that of the xanthophyll lutein (28), the primary pigment present in etiolated cucumber cotyledons (29). Other carotenoids (e.g., carotene, probably β-carotene, and one or more xanthophylls) are also present (unpublished results).

Obtaining a well executed action spectrum should make it possible to match wavelengths most efficient in initiating the herbicidal response with those of a specific plant pigment.

Results from chlorophyllous mutants of rice (*Oryza sativa* L.) (5), corn, and soybeans (*Glycine max* L. Merr.) (4), support the contention that a carotenoid is involved in activating the DPE molecule. Yellow and green mutants were equally susceptible to herbicidal injury; however, the albino mutants were resistant. Cucumber seedlings pretreated with fluridone, a known carotenoid biosynthesis inhibitor (30), were also resistant (3). By quantitatively examining herbicidal activity differences in tissues treated with compounds (e.g., CPTA [2-(4-chlorophenylthio)-triethylamine hydrochloride], onium compounds, etc.) capable of regulating the relative amounts of carotenoids present (31, 32, 33), the pigment(s) involved in the light-activating mechanism may be identified. Algal carotenoid mutants or various *Neurospora* species known to have specific carotenoids may also aid in the study of DPE-pigment interactions.

DPE-Carotenoid Interactions in Vitro. Observations of spectral changes occurring upon illumination of DPE-treated crude pigment extracts have been attempted (34, 35). These experiments were successful only with nitrofen (35). Nitrofen attacks the 4th free ring of the chlorophyll molecule in methanol and carbon tetrachloride, thus decreasing absorbance in the visible region (35). There was an apparent association between the solute and solvent molecules in pyridine that effectively slowed the reduction in absorbance. Because light-activation of the DPE molecule *in vivo* does not require chlorophyll, similar studies were conducted with carotenoids (unpublished results, 34). To date, none of these experiments have been successful. Protein-carotenoid complexes destroyed by the organic extraction procedures used might be required for herbicidal activation. For carotenoids to function properly they may exist as a protein-pigment complex imbedded within the hydrophobic matrix of cell membranes. The involvement of this complex is partly based on the difficulty in demonstrating carotenoid fluorescence (37). One of the functions of carotenoids in photosynthesis is the transfer of energy to chlorophylls. Perhaps, DPE's directly or indirectly intercept the electron or energy transfers normally occurring between carotenoids and chlorophylls (38, 39, 40).

The outer chloroplast and etioplast envelope contains carotenoids (36). By illuminating this subcellular fraction at various wavelengths in a spectrophotometer and obtaining a difference spectra in the presence of DPE herbicides, the direct interaction between herbicide and pigment can be evaluated. The wavelength at which the interaction is observed will implicate the pigment involved.

Carotenoids vs. Flavins. To date, considerable evidence supports the contention that carotenoids are involved in the light-activating mechanism of DPE's (3, 4, 5, 7-10). From a purely physical view, a flavin can be perceived as a more likely candidate for activation of DPE's than can a carotenoid. The following molecular properties that favor riboflavin over β-carotene are from a review by Song, Moore, and Sun (37): (a) Riboflavin is capable of intensely fluorescing, whereas the fluorescence of β-carotene is weak and anomolous. (b) Riboflavin will phosphoresce in the visible region, whereas β-carotene will not. (c) (n, π*) states are available for riboflavin but, not for β-carotene. (d) Riboflavin will decompose to yield photoproducts and can also initiate various intermolecular photooxidations. β-carotene can undergo cis ⇄ trans photoisomerizations but is incapable of intermolecular photooxidations. (e) Riboflavin will generate singlet oxygen by triplet energy transfer and yield hydrogen peroxide with subsequent restoration of flavin. β-carotene quenches singlet oxygen and carotene is consumed. (f) Riboflavin has the photochemical ability to form radicals, whereas β-carotene probably does not. Examination of effects of DPE's on tissues [e.g., cucumber hypocotyl (41)] with relatively high riboflavin to carotenoid ratios should be informative.

Spin-trapping Experiments. The paraquat (1,1'-dimethyl-4, 4'-bipyridinium ion) radical generated following reduction by PS I yields superoxide anions in the presence of oxygen (2). The superoxide radical was detected with spin-trapping techniques and ESR spectroscopy (42). In these experiments, isolated chloroplasts were illuminated in the presence of a spin adduct and paraquat. AFM-induced injury also appears to require oxygen. However, pretreatment of chloroplasts with DABCO [1,4-diazobicyclo (2,2,2)-octane], a widely used quenching agent of singlet oxygen did not prevent damage by oxyfluorfen (43). Although superoxide may not be involved in DPE injury, these same spin-trapping techniques may prove useful in identifying the types of reactions occurring in illuminated chloroplasts or etioplasts treated with DPE's.

Reaction Mechanism(s) of DPE's. DPE's could potentially be degraded during light-activation reactions or radical reaction sequences. Identification of DPE breakdown products from susceptible plants kept in the dark or in the light would test this possibility. If the products obtained under these two conditions are different, identification of these products may assist in identifying the light-activation mechanism.

A free radical mechanism was proposed for the expression of phytoxicity by ioxynil (4-hydroxy-3,5-diiodobenzonitrile) (44). Ioxynil undergoes radical reactions with benzene upon illumination with UV light. Identification of the reaction products by thin-layer chromatography determined the types of radical reactions occurring under these conditions. Similar experiments may yield information concerning light reactions of DPE's *in vivo*.

Injury to Cell Membranes

According to the fluid-mosaic model of membrane structure (45), cell membranes consist of a fluid phospholipid bilayer. Embedded within this bilayer are globular proteins essential to membrane function. A large class of phospholipids present in membranes are phosphoglycerides (15). The fatty acid in the number 2 position is often unsaturated (46). In plants, the unsaturated fatty acid is frequently linolenic acid (15); with 3 double bonds ($18{:}3^{\Delta 9,\ 12,\ 15}$).

Lipophilic Free Radical Reactions. The divinyl methane structure present in PUFA is susceptible to hydrogen abstraction with subsequent formation of a fairly stable free radical (47). In the presence of the proper initiation factors, these reactions can be induced within the hydrophobic matrix of the membrane (48). Once these reactions have started, there can be considerable cell damage (49). The orderly array of fatty acids present in membranes permits maximum interaction of the individual molecules and thus, readily propagates free radical reactions.

Propagation reactions occur to the greatest extent when oxygen is abundant (48); however, the rates of such reactions are proportional to oxygen content only at low partial pressures. To test whether DPE's might be involved in initiating propagation reactions, activity of AFM was examined in tissues kept in an atmosphere of either oxygen or nitrogen (3). The activity of AFM was significantly reduced when tissues were maintained in a nitrogen atmosphere. A related compound, fluorodifen, was also found to require oxygen and light for maximum activity (21).

Lipid Analyses. Verification of PUFA radical chain reactions induced by light-activated DPE molecules can be made by examining changes in a polar lipid fraction collected from injured tissues. Radical chain reactions can be terminated through cross-reactions of the endoperoxides formed (48). Therefore, either an appearance of lipid polymers or disappearance of certain phospholipids should occur. If only the PUFA moieties are involved in these reactions, a loss of PUFA in the DPE-treated tissue or an increase in the ratio of saturated to unsaturated fatty acids should occur.

Electron micrographs from cucumber cotyledons reveal the presence of large numbers of lipid bodies (oleosomes). Lipids in oleosomes are comprised mostly, if not entirely, of triacylglycerols (46). When studying DPE injury symptoms it is important to examine changes in a polar lipid fraction and not a total lipid extract. The reasons are threefold. First, if the primary site of action of DPE's resides in cell membranes, it is logical to look for changes only in the structural components of those membranes; the polar phospholipids. Second, interactions of the light-activated DPE molecule with PUFA's in lipid bodies would probably not result in any real cellular damage. Preliminary experiments with AFM indicate there is little interaction between the herbicide and fats of oleosomes (unpublished results). Third and most important, the high quantities of neutral lipids in oleosomes may mask the detection of any herbicide effect on membrane lipids. Even if significant changes in phospholipids cannot be detected, the possibility of radical reactions being initiated as a result of DPE-treatment cannot be discounted. Small perturbations in the fatty acid moieties would likely result in large changes in membrane permeability characteristics observed. This would result in cellular decompartmentalization and death.

Products of Lipid Peroxide Decomposition. In view of the potential problems with direct lipid analysis, a simpler and more sensitive assay, the detection of thiobarbituric acid-reacting materials (TBARM), was used to detect DPE injury to membranes (3). Various non-volatile precursors of malonyl dialdehyde (MDA) are products of lipid peroxide decomposition *in vitro* and *in vivo* (48). These products result from free radical chain reactions involving plant PUFA with three double bonds; e.g. linolenic acid (50). The MDA-precursor materials can be detected using a colorimetric reaction with thiobarbituric acid (TBA) (50-53). Detection of TBARM following the light-activation of AFM in treated cucumber cotyledons indicates there is direct physical damage to the membranes (3). Plasma membrane, tonoplast, and chloroplast envelope disruption has been verified by electron microscopy (Figure 3). More important, the presence of TBARM in damaged tissue provides the first real evidence that injury to the membranes results from the formation of highly reactive and destructive lipophilic free radicals (51, 52, 54-58). Because fluorescence measures various products of lipid peroxidation (59), this technique could be used to further examine DPE-induced injury to cell membranes.

The evolution of short chain hydrocarbon gases (SCHG) can also be used as an indication of lipid peroxidation (54, 60, 61, 62). Kunert and Böger (43) detected ethane evolution from oxyfluorfen-treated *Scenedesmus* cells within 1 to 3 h. Isolated chloroplasts evolve ethane within 15 to 30 min following treatment and light-activation of oxyfluorfen. Cucumber cotyledons treated with 1 μM AFM for 6 h in darkness showed a significant increase in the amount of ethylene produced 30 min after exposure to light (unpublished results). Longer chain hydrocarbons (e.g., pentane) have not been detected (unpublished

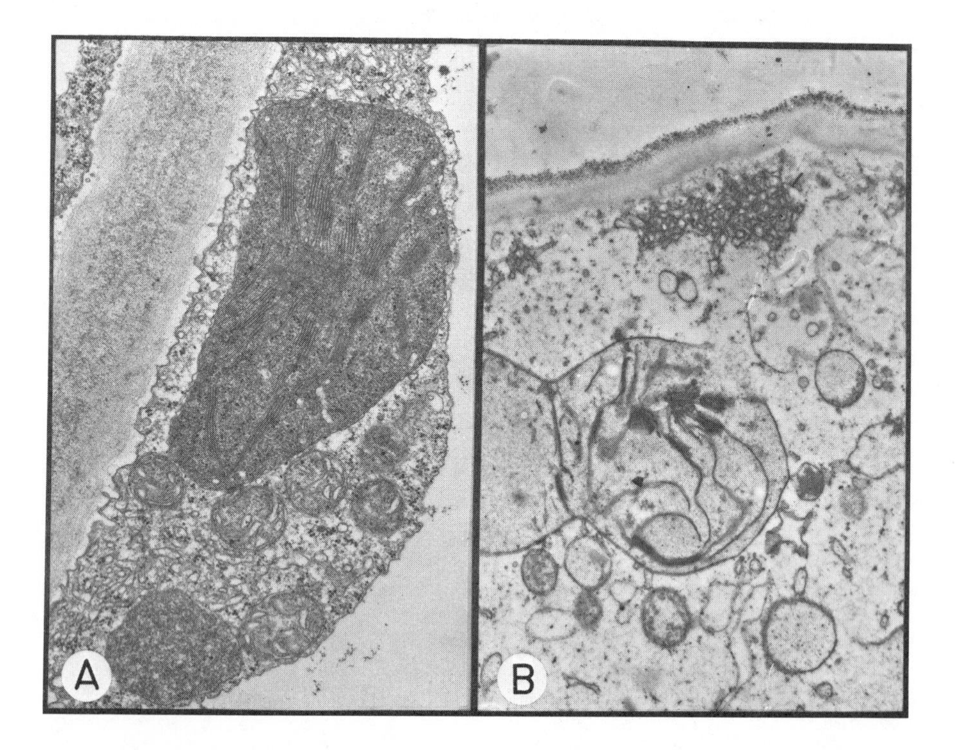

Figure 3. Electron micrographs of cells from greened cucumber cotyledons pretreated with 1 μM AFM for 6 h prior to light (600 μE/m² s, PAR) exposure.

A, Ultrastructure prior to light exposure (3). Glutaraldehyde plus osmium tetroxide fixation. Magnification = 20,000×.

B, Cellular disruption after a 45-min light exposure. Potassium permanganate fixation. Magnification = 8,000×.

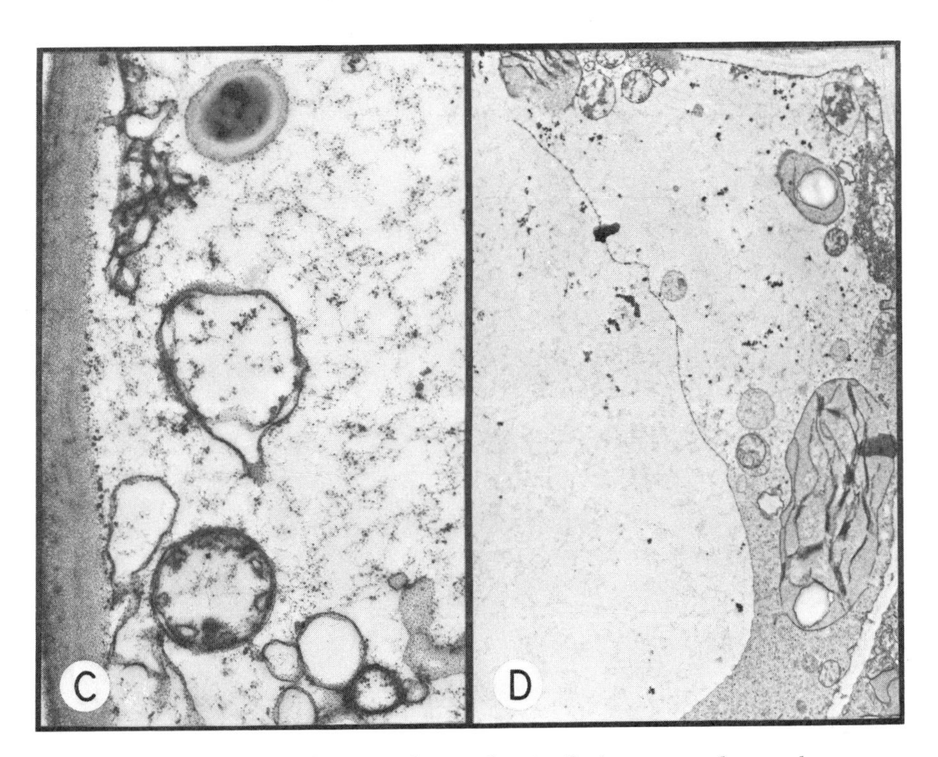

Figure 3. (continued) Electron micrographs of cells from greened cucumber cotyledons pretreated with 1 μM AFM for 6 h prior to light (600 μE/m² s, PAR) exposure.

C, Vesiculation of plasmalemma after a 60-min light exposure. Glutaraldehyde plus osmium tetroxide fixation. Magnification = 21,000×.

D, Cytoplasmic damage following tonoplast disruption. Potassium permanganate fixation. Magnification = 5,000×.

results, 43). TBARM detected in AFM-treated cucumber cotyledons supports the contention that SCHG evolved from these tissues was the result of direct interaction of herbicide and PUFA, and not merely an indication of cellular death.

Spin-trapping Experiments. As mentioned earlier, some of the radical chain reactions initiated by light-activated DPE's can terminate with the subsequent formation of PUFA polymers (48). Spin-trapping techniques have been used to study reactions of lipoxygenase with linoleic acid (63). It was proposed that the formation of dimeric linoleic acid required the involvement of hydroperoxylinoleic acid (64). The ESR spectrum from the interactions of the PUFA radical and the spin-trap indicated this involvement occurred. Similar spin-trapping techniques could be used to investigate the possibility that DPE's may ultimately induce formation of various fatty acid dimers (e.g., dimers of linolenic acid). Formation of such polymers should dramatically affect membrane function.

Involvement of Polyunsaturated Fatty Acids. To further test the importance of PUFA, the relative activity of DPE's following tissue pretreatment with drugs (e.g., substituted pyridazinones) known to change the degree of unsaturation within the membrane (65) should be studied. Compounds increasing the degree of saturation might afford the tissue some protection against DPE injury, whereas, compounds increasing the degree of unsaturation should make the tissue more susceptible.

The relative amounts of saturated and unsaturated fatty acids in the membrane can be altered by maintaining tissue in different temperature regimes (46). Organisms growing in cool climates have increased levels of unsaturated fatty acids in their membranes. As a result, their membranes are more fluid (the melting point of phospholipids decreases with increasing numbers of unsaturated fatty acid components). These tissues should be more susceptible to DPE-injury.

Etiolated cucumber cotyledons showed dramatic increases in fatty acid desaturase activity following exposure to light (66). Consequently, green cotyledons have significantly more unsaturated fatty acids than etiolated cotyledons. However, AFM appears to be as active in etiolated tissue as in green tissues (3). This can be explained by increased levels of radical scavengers in greened tissue.

Effects of Antioxidants. The most important evidence implicating free radicals in DPE membrane injury is the ability to protect against damage with a known radical scavenger. Compounds with the potential ability to provide limited protection against DPE injury by free radical scavenging and other mechanisms include BHA (butylated hydroxyanisole), BHT (butylated hydroxytoluene), EDU {N-[2-(2-oxo-1-imidazolidinyl)-

ethyl]-N'-phenylurea}, DPPD (N,N'-dimethyl-p-phenylenediamine), DABCO, diphenylamine, sodium benzoate, SOD (superoxide dismutase) and α-T. However, except for α-T (Figure 4), the compounds tested have afforded little or no protection (unpublished data, 43). The inability of DABCO and SOD to reduce oxyfluorfen injury indicates singlet oxygen is not the destructive agent (43). However, the ability of α-T, a known *in vivo* scavenger of lipophilic free radicals, to protect against AFM injury suggests the herbicide initiates a free radical chain reaction with PUFA moieties (e.g., linolenic acid) of the phospholipid molecules making up cell membranes (3).

In our experiments, limited protection of AFM-induced injury to cucumber cotyledons was obtained with BHA and BHT. These two compounds showed some synergistic characteristics. However, the concentrations necessary for protection were high (400 μM) and caused some injury to the controls. The concentrations of BHA and BHT needed to protect the tissue should be higher than for α-T. BHA and BHT each have an antioxidant stoichiometric factor (n) equal to 1, whereas for α-T, n can be equal to 2 (67). An n of 2 for α-T means this molecule has the ability to quench up to 2 radicals before being destroyed. DPPD has been reported to have antioxidant activity (68). However, in our tests DPPD was toxic at high concentrations (400 μM) and ineffective at lower concentrations (e.g., 50 μM).

In Vitro Assays. The proposed DPE mechanism of initiating and propagating radical chain reactions should be investigated. For example, the effects of "acid synergists", such as EDTA (ethylenediaminetetraacetic acid), citrate, and ascorbate, should be examined with respect to their ability to decrease the rate of lipid oxidation by preventing radical initiation or propagation (48). After inducing injury in herbicide-treated tissues, one could study the effects of transition metals (e.g., Cu or Fe) on the metal-catalyzed decomposition of the hydroperoxides formed. Kunert and Böger (43) detected ethane evolution from oxyfluorfen-treated chloroplasts in the presence of Fe-EDTA. However, the addition of ferredoxin to this fraction instead of Fe-EDTA yielded only 20% of the ethane evolved with iron present.

Further studies on the more intricate details of DPE-induced radical initiation and propagation must await the development of an *in vitro* assay sensitive to these herbicides. For example, using various subcellular fractionation techniques, all of the separated cellular components required for expression of herbicidal activity can be recombined. Injury may occur by combining an outer etioplast envelope fraction (the source of carotenoids) with plasma membrane (or microsomal fraction) obtained from oat (*Avena sativa* L.) roots, and exposing them to light and herbicide. DPE-induced injury could be monitored by many of the standard methods for determining autoxidation of lipids (e.g., detection of TBARM, evolution of SCHG, consumption

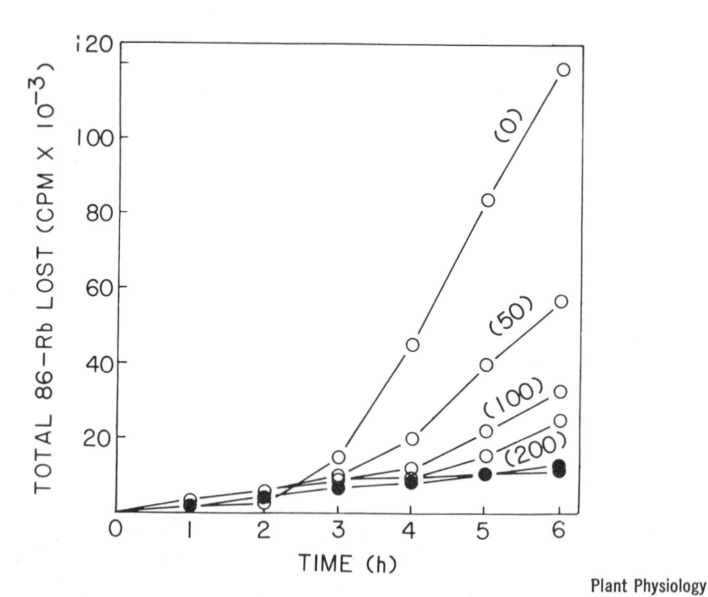

Figure 4. Efflux of $^{86}Rb^{+}$ from cucumber cotyledons treated with 1 μM AFM in the presence of various concentrations (0, 50, 100, and 200 μM) of α-tocopherol (3). At time zero, cotyledons were exposed to herbicide and α-tocopherol in the dark in a nitrogen atmosphere. After 1 h, the atmosphere was changed to air and at 2 h the cotyledons were exposed to light (600 μE/m² s, PAR). Closed circles are effluxes from control tissue treated with 1.0% ethanol or 1.0% ethanol plus 200 μM α-tocopherol.

of O_2, or measurement of various fluorescent products of lipid peroxidation). In this particular system, it may be possible to detect injury by measuring activity of the enzyme marker for oat root plasma membrane (69); K^+-stimulated MgATPase.

Another *in vitro* system for studying details of the mechanism of action of DPE's involves the binding of commercially available PUFA (e.g., linolenic acid) to silica plates (48) and adding the various subcellular components thought to be required for herbicidal activity. The DPE-initiated radical reactions in the PUFA monolayers could be followed by detection of TBARM or by measurement of oxygen consumption (59).

Cellular Compartmentation of Herbicides. For detailed mechanism of action research, the *in situ* tissue distribution of DPE's should be known. One method to determine location is to treat with radiolabeled herbicide and, at some later time, fractionate the tissue into its various subcellular components (70). The relative amount of radioactivity present in the various organelles indicates the cellular location of the herbicide. There are, however, contamination problems associated with this technique.

An alternative is to utilize the technique developed by MacRobbie and Dainty (71), and Pitman (72), for intracellular location of ions: (i.e., compartmental analysis). The cellular compartment containing herbicide can be determined by loading the tissue with radiolabeled herbicide for relatively long periods of time and then studying the kinetics of efflux once the external label is removed. This technique has been used by Price and Balke (73) to determine the cellular compartmentation of ^{14}C-atrazine [2-chloro-4-(ethylamino)-6-(isopropylamino)-*s*-triazine]. Compartmental analysis assumes tissue is in a steady-state equilibrium with the radioactive label. The analysis would be invalid if the herbicide was exerting its effect during the experiment. Because DPE's require light for activity, this problem can be circumvented by doing the analysis in darkness.

Ultrastructural Analyses

Ultrastructural analyses have been conducted to identify structural changes in the cellular membrane system of green and etiolated cucumber cotyledons treated with AFM (3, 74). Prior to preparation for microscopy, cotyledons treated with 1 μM AFM for 6 h in the dark were exposed to high light (600 $\mu E\ m^{-2}\ s^{-1}$) for periods up to 1 h.

The ultrastructure of the tissue treated with AFM for 6 h in darkness (Figure 3A) was the same as untreated tissue. However, massive cellular and membrane damage was apparent in the AFM-treated tissue within 30 to 45 min following light-activation of the herbicide (Figure 3B). Some etioplasts and chloroplasts were swollen, which is characteristic of an uncoupled organelle; i.e., the organelle is indiscriminately permeable to solutes (in particular, protons) (75). Very often, large holes occurred in the plasma membrane, tonoplast, and some chloroplasts and etioplasts. Also apparent were invaginations of membranes and, revesiculation of the invaginations once they had broken away from the continuous membrane system (Figure 3C). Most of the other cellular organelles were severely damaged or destroyed.

Early signs of injury appeared in the outer chloroplast or etioplast envelope and in the tonoplast. This information supports the proposed mechanism of action of DPE herbicides. Carotenoids are known to be present in the outer plastid envelope (36), and there is also a substantial amount of linolenic acid esterifed to the galactolipids and phospholipids of this membrane (15). Interestingly, there was only limited observable damage to the thylakoids of tissue exposed to light for short periods of time. This is reasonable because tocopherols and tocopherylquinones are located in the thylakoids (15).

A secondary effect in the sequence of events leading to cellular death results from the disruption of the tonoplast. The release of various hydrolytic enzymes from the vacuole is extremely detrimental to the surrounding cytoplasm (Figure 3D).

Summary

A model of the proposed mechanism of action of a DPE, such as AFM, is outlined diagrammatically in Figure 5. Light absorbed by yellow plant pigments (carotenoids) activates the AFM molecule. The carotenoid molecule involved appears to be destroyed following the activation of the herbicide. The light-activated form of the AFM molecule is then involved, either directly or indirectly, in the initiation of a radical chain reaction through the abstraction of a hydrogen atom from the divinyl methane structure in PUFA. This relatively stable free radical subsequently reacts with molecular oxygen to form a lipid peroxide that readily propagates throughout the hydrophobic matrix of the membrane.

The propagation reactions can be terminated in a number of ways. One termination sequence involves cross reactions of fatty acid moieties resulting in the formation of polymers (48). The formation of these polymers would profoundly affect the fluidity and, therefore, the permeability characteristics of the membrane. The propagation reactions could also be terminated through decomposition of the lipid peroxides formed, resulting in a physical disintegration of some portions of the membrane. Some of the products of lipid peroxide decompositon (e.g., MDA) are known to undergo Schiff's-base-type reactions with the amino acids of proteins (76). However, the detrimental effect of covalently cross-linking proteins is probably secondary in nature. The radical chain can also be terminated in a

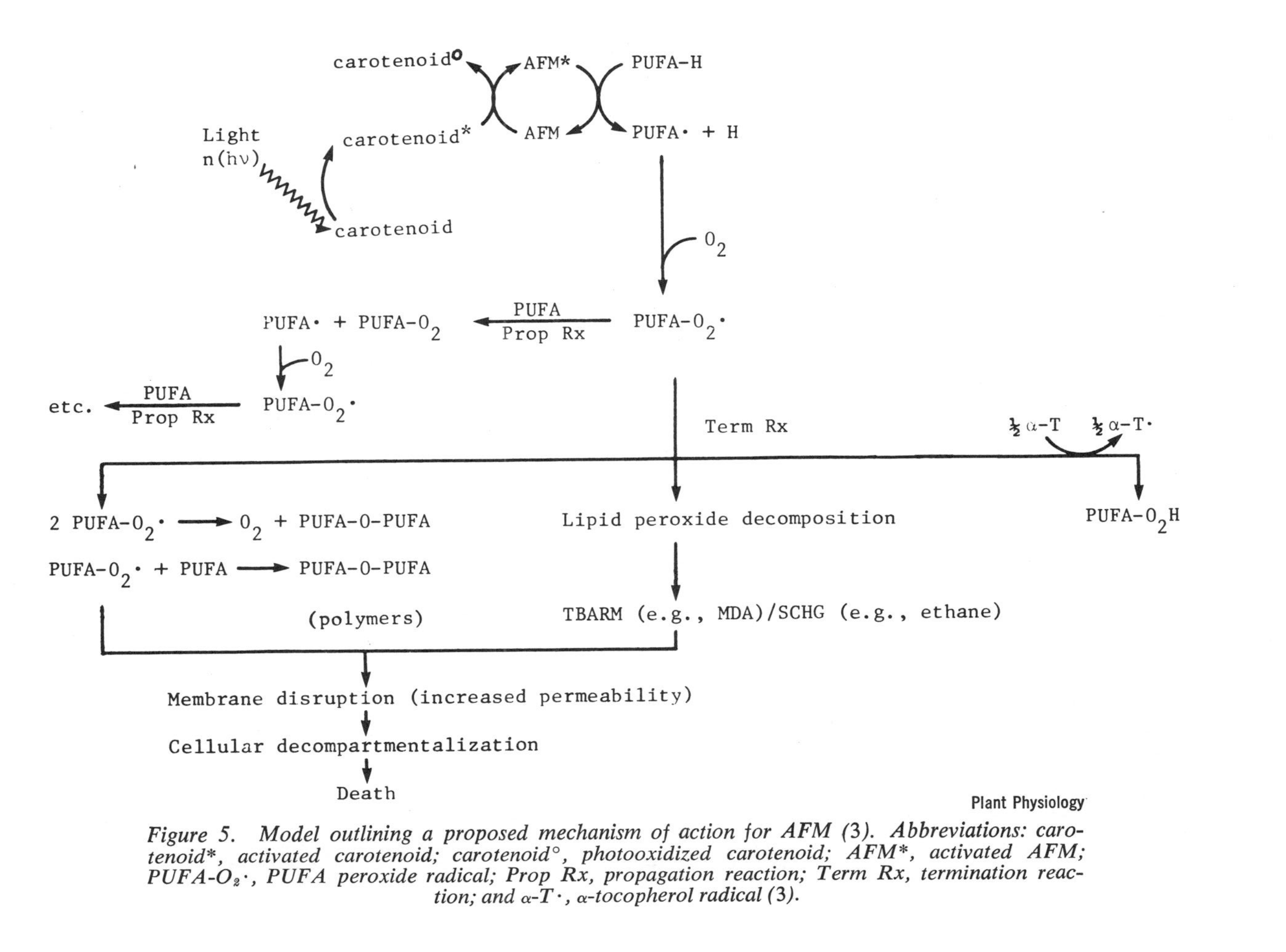

Figure 5. Model outlining a proposed mechanism of action for AFM (3). Abbreviations: carotenoid, activated carotenoid; carotenoid°, photooxidized carotenoid; AFM*, activated AFM; PUFA-O_2·, PUFA peroxide radical; Prop Rx, propagation reaction; Term Rx, termination reaction; and α-T·, α-tocopherol radical (3).*

nondestructive manner through a competitive antioxidant reaction involving a scavenger of lipophilic free radicals (61).

Acknowledgements

This work was supported by a grant from Mobil Foundation and Mobil Chemical Company.

Literature Cited

1. Orr, G. L.; Hess, F. D. Pestic. Biochem. Physiol. In Press.
2. Evans, C. A. Aldrichimica Acta 1979, 50, 23-9.
3. Orr, G. L.; Hess, F. D. Plant Physiol. In Press.
4. Fadayomi, R. O.; Warren, G. F. Weed Sci. 1976, 24, 598-600.
5. Matsunaka, S. J. Agr. Food Chem. 1969, 17, 171-5.
6. Matsunaka, S. Residue Rev. 1969, 25, 45-8.
7. Matsunaka, S. JAQR 1972, 6, 189-94.
8. Prendeville, G. N.; Warren, G. F. Weed Res. 1977, 17, 251-8.
9. Pritchard, M. K.; Warren, G. F.; Dilley, R. A. Weed Sci. 1980. 28, 640-5.
10. Vanstone, D. E.; Stobbe, E. H. Weed Sci. 1979, 27, 88-91.
11. William, R. D.; Warren, G. F. Weed Res. 1975, 15, 285-90.
12. Yih, R. Y.; Swithenbank, C. J. Agr. Food Chem. 1975, 23, 592-3.
13. Pereira, J. F.; Splittstoesser, W. E.; Hopen, H. J. Weed Sci. 1971, 19, 647-51.
14. Naylor, A. W. "Herbicides: Physiology, Biochemistry, Ecology", Vol. 1, Audus, L. J., Ed.; Academic Press: New York, 1976; pp. 397-426.
15. Goodwin, T. W.; Mercer, E. I. "Introduction to Plant Biochemistry"; Pergamon Press: New York, 1972; 356 pp.
16. Boguth, W.; Repges, R. Int. J. Vit. Nutr. Res. 1969, 39, 289-95.
17. Hess, F. D. Weed Sci. 1980, 28, 515-20.
18. Mino, M. J. Nutr. Sci. Vitaminol. 1973, 19, 95-104.
19. Summers, L. A. "The Bipyridinium Herbicides"; Academic Press: New York, 1980; 449 pp.
20. Scott, M. L. "Handbook of Lipid Research 2. The Fat-Soluble Vitamins", DuLuca, H. F., Ed.; Plenum Press: New York, 1978; pp. 133-210.
21. Van Assche, C. J.; Ebert, E. Compte rendu 6e Conference du Comite Francais de Lutte contre les Mauvaises Herbes (Columbia), 1971; 15-31.
22. Bugg, M. W.; Whitmarsh, J.; Reick, C. E.; Cohen, W. S. Plant Physiol. 1980, 65, 47-50.
23. Eastin, E. F. Abstr. Weed Sci. Soc. Amer. 1972, No. 93.

24. Kruger, P. S.; Ward, D.; Theissen, R. J.; Downing, C. R.; Kaufman, H. A. Proc. 12th Brit. Weed Contr. Conf. 1974, 839-45.
25. Moreland, D. E. "Progress in Photosynthesis Research", Metzner, H., Ed.; Int. Union Biol. Sci.: Tubingen, 1969; pp. 1693-1711.
26. Moreland, D. E.; Blackmon, W. J.; Todd, H. G.; Farmer, F. S. Weed Sci. 1970, 18, 636-42.
27. Pollak, T. Diss. Abstr. Int. B Sci. Eng. 1974, 34, 5752.
28. Zscheile, F. P.; White Jr., J. W.; Beadle, B. W.; Roach, J. R. Plant Physiol. 1942, 17, 331-46.
29. Rebeiz, T. R. Inst. Rech. Agron. Liban. 1968, No. 21.
30. Bartels, P. G.; Watson, C. W. Weed Sci. 1978, 26, 198-203.
31. Simpson, D. J.; Chichester, C. O.; Lee, T. H. Aust. J. Plant Physiol. 1974, 1, 119-33.
32. Simpson, D. J.; Rahman, F. M. M.; Buckle, K. A.; Lee, T. H. Aust. J. Plant Physiol. 1974, 1, 135-47.
33. Yokoyama, H.; Hsu, W. J.; Poling, S. M.; Hayman, E. Chapter 9 in this book.
34. Pritchard, M. K. "Plant Response to Light-induced Herbicides"; Ph.D. Thesis; Purdue University, West Lafayette, IN, 1979; 83 pp.
35. Parthasarathy, S.; Purushothaman, D. Indian J. Agr. Chem. 1977, 8, 247-51.
36. Douce, R.; Joyard, J. "Advances in Botanical Research", Vol. 7, Woolhouse, H. W., Ed.; Academic Press: New York, 1979; pp. 1-116.
37. Song, P. S.; Moore, T. A.; Sun, M. "The Chemistry of Plant Pigments", Chichester, C. O., Ed.; Academic Press: New York, 1972; pp. 33-74.
38. Nobel, P. "Biophysical Plant Physiology"; W. H. Freeman and Co.: San Francisco, 1974; 488 pp.
39. Aquist, G.; Samuelsson, G.; Bishop, N. I. Physiol. Plant. 50, 63-70.
40. Sauer, K. "Bioenergetics of Photosynthesis", Govindjee, Ed.,; Academic Press: New York, 1975; 698 pp.
41. Hertel, R.; Jesaitis, A. J.; Dohrmann, U.; Briggs, W. R. Planta 1980, 147, 312-19.
42. Horbaum, J. R.; Bolton, J. R. Biochem. Biophys. Res. Commun. 1975, 64, 803-5.
43. Kunert, K. J.; Boger, P. Weed Sci. 1981, 29, 169-73.
44. Wain, R. L. Proc. 7th Br. Weed Control Confr. 1964, 306-11.
45. Singer, S. J.; Nicolson, G. L. Science 1972, 175, 720-31.
46. Lehninger, A. L. "Biochemistry: The Molecular Basis of Cell Structure and Function"; Worth Publishers: Inc., New York, 1975; 1104 pp.
47. Uri, N. "Autoxidation and Autioxidants", Vol. I, Lundberg, W., Ed.; Wiley (Interscience): New York, 1961; pp. 55-106.

48. Mead, J. F. "Free Radicals in Biology", Vol. I, Pryor W., Ed.; Academic Press: New York, 1976; pp. 51-68.
49. Green, J. "The Fat-Soluble Vitamins", DeLuca, H.; Suttie, J., Eds.; The University of Wisconsin Press: Madison, 1969; pp. 293-347.
50. Dahle, L. K.; Hill, E. G.; Holman, R. T. Arch. Biochem. Biophys. 1962, 98, 253-61.
51. Kwan, T. W.; Olcott, H. S. Nature 1966, 210, 214-5.
52. Niehaus, W. G.; Samuelson, B. Eur. J. Biochem. 1968, 6, 126-30.
53. Placer, Z. A.; Cushman, L. L.; Johnson, B. C. Anal. Biochem. 1966, 16, 359-64.
54. Lizada, C. C.; Yang, S. F. Plant Physiol. 1980, 65S, 802.
55. McKnight, R. C.; Hunter Jr., F. E. Biochim. Biophys. Acta 1965, 98, 640-6.
56. Ottolenghi, A. Arch. Biochem. Biophys. 1959, 79, 355-63.
57. Pryor, W. A.; Stanley, J. P. J. Org. Chem. 1975, 40, 3615-7.
58. Wills, E. D.; Wilkinson, A. E. Biochem. J. 1966, 99, 657-66.
59. Tappel, A. L. "Free Radicals in Biology", Vol. IV, Pryor, W., Ed.; Academic Press: New York, 1980; pp. 1-47.
60. Dumelin, E. E.; Tappel. A. L. Lipids 1977, 12, 894-9.
61. McCoy, P. B.; Pfeifer, P. M.; Stipe, W. H. Ann. NY Acad. Sci. 1972, 203, 62-73.
62. Simcox, D. P.; Ku, H. S. Plant Physiol. 1980, 65S, 223.
63. deGrott, J. J. M. C.; Garssen, G. J.; Vliegenthart, J. F. G.; Boldingh, J. Biochim. Biophys. Acta 1973, 326, 297-304.
64. Garssen, G. J.; Vliegenthart, J. F. G.; Boldingh, J. Biochem. J. 1971, 122, 327-33.
65. St. John, J. B. Chapter 6 in this book.
66. Murphy, D. J.; Stumpf, P. K. Plant Physiol. 1979, 63, 328-35.
67. Reich, L.; Stivala, S. S. "Autoxidations of Hydrocarbons and Polyolefins"; Dekker: New York, 1969; 154 pp.
68. Wasserman, R. H.; Taylor, A. W. Annu. Rev. Biochem. 1972, 41, 149-206.
69. Hodges, T. K.; Leonard, R. T. Methods Enzymol. 1974, 33, 392-406.
70. Boulware, M. A.; Camper, N. D. Weed Sci. 1973, 21, 145-59.
71. MacRobbie, E. A. C.; Dainty, J. Physiol. Plant. 1958, 11, 782-801.
72. Pitman, M. G. Aust. J. Biol. Sci. 1963, 16, 647-68.
73. Price, T. P.; Balke, N. E. Abstr. Weed Sci. Soc. Amer. 1981, No. 203.
74. Hess, F. D.; Orr, G. L. Abstr. Weed Sci. Soc. Amer. 1981, No. 249.
75. Poole, R. J. Annu. Rev. Plant Physiol. 1978, 29, 437-60.
76. Chio, K. S.; Tappel, A. L. Biochemistry 1969, 8, 2827-32.

RECEIVED September 24, 1981.

9

Bioregulation of Pigment Biosynthesis by Onium Compounds

H. YOKOYAMA, W. J. HSU, S. POLING, and E. HAYMAN

U.S. Department of Agriculture, Agricultural Research Service,
Fruit and Vegetable Chemistry Laboratory, Pasadena, CA 91106

A large number of onium compounds regulate carotenoid pigment biosynthesis in plant tissues. They stimulate carotenoid biosynthetic pathways with no apparent herbicidal activity. A substantial amount of information on structure-activity relationships for bioregulation of carotenoids in higher plants has been developed. Both the *trans*- and *cis*- biosynthetic systems can be regulated. The carotenoid pattern observed is determined essentially by the nature of onium compounds employed. For example, 2-diethylaminoethyl-4-methylphenylether caused a large accumulation of all *trans*-lycopene (ψ,ψ-carotene), whereas 2-diethylaminoethylhexanoate induced the formation of β-carotene (β,β-carotene). The compound *N*-methyl,*N*-hexylbenzylamine stimulated the formation of *cis*-carotenes.

Regulation of the biosynthesis of the carotenoids was demonstrated in the early 1950's when it was shown that β-ionone stimulated carotenogenesis in the fungus *Phycomyces blakesleeanus* (1, 2, 3) without itself being incorporated into the carotene molecules (4). A similar stimulatory effect of β-ionone was observed on carotene production in heterothallic cultures of *Phycomyces blakesleeanus* (5). In *Blakeslea trispora*, an effect somewhat similar to that observed with β-ionone was seen with trisporic acid (6). Trisporic acid only stimulated carotenogenesis in the (-) strain (7). Both β-ionone and trisporic acid do not appear to affect carotene biosynthesis in higher plants (8). Ninet *et al*. (9) investigated the biosynthesis of carotenoids by *B. trispora* in the presence of various nitrogenous compounds. Pyridine, imidazole, and some of their derivatives were found to stimulate the synthesis of *trans*-lycopene (ψ,ψ-carotene), whereas isonicotinoylhydrazine, succinimide, and 4-hydroxypyridine enhanced β-carotene (β,β-carotene) synthesis.

In 1969, Knypl (10) found that an onium compound, chlormequat [(2-chloroethyl)-trimethylammonium chloride], caused all-*trans*-lycopene accumulation in detached pumpkin (*Curcubita pepo*) cotyledons. However, no such response was observed when chlormequat was applied to other plant tissues (11). A major step forward was observed in 1970 with the discovery that another onium compound, 2-(4-chlorophenylthio)-triethylammonium chloride (CPTA), can cause the accumulation of lycopene in a wide array of plant tissues and microorganisms (11).

Since the discovery of CPTA, research on onium compounds has generated a substantial amount of information relative to the structure-activity relationships for bioregulation of carotenoid pigments in higher plants and microorganisms. Both the *trans*- and *cis*-biosynthetic pathways can be regulated. The carotenoid pattern observed is determined essentially by the nature of the onium compounds employed.

All-trans Carotenes

General Formula. In the regulation of the *trans*-carotenogenic system, the onium compounds have the special formula:

$$(C_2H_5)_2N-CH_2-R$$

In general, the *N,N*-dimethyl analogs are less effective than the *N,N*-diethyl compounds in promoting carotenoid formation. The magnitude of the stimulation, but not the general pattern of carotenoid response, depends on R (12). The nature of R can cause modifications in the amount of the individual carotenoids. Thus, when R lacks an aromatic ring, there appears to be less inhibition of the cyclase(s); consequently, more cyclic carotenoids accumulate. Examples are given in Tables I-III. All of the compounds tested caused lycopene accumulation in Marsh grapefruit (*Citrus paradisi*). The untreated fruit had the normal light yellow color. After treatment, the color of the flavedo ranged from light orange to an intense red. The flavedo of all treated fruit showed lycopene accumulation (Tables I, II, and III). Lycopene was not detected in the untreated fruits and it is not normally present in mature grapefruit (13). The biological activity was also correlated with the logarithm of the 1-octanol/water partition coefficient (log *P*).

Diethylalkylamines. Treatments with different diethylalkylamines (Table I) gave a fairly consistent response pattern as the length of the alkyl group was increased. The amount of any given carotene remained about the same or increased slightly with

Table I. Effect of trialkylamines on carotene content of the flavedo of Marsh seedless grapefruit (μg/g dry wt). Postharvest treatment of fruits, and method of isolation and identification of pigments are described in reference 12. A portion of ground flavedo was dried *in vacuo* at 65 C to obtain the dry wt.

	Control	Compound*				
		1	2	3	4	5
Phytofluene	37.3	38.6	29.0	28.3	27.8	39.5
ζ-Carotene	2.25	2.47	3.13	4.32	14.6	17.8
Neurosporene	1.72	1.28	1.37	1.38	2.94	5.16
Lycopene		1.01	6.99	59.0	143	115
γ-Carotene	0.37	0.30	0.96	1.17	2.59	3.21
α-Carotene	0.54	1.04	1.02	1.26	2.07	2.11
β-Carotene	1.72	1.41	1.20	1.35	4.73	6.75
Total	43.9	46.1	43.7	96.8	197.7	189.5
Log P		2.94	3.44	3.94	4.44	4.94

*1. $(C_2H_5)_2N(CH_2)_4CH_3$
2. $(C_2H_5)_2N(CH_2)_5CH_3$
3. $(C_2H_5)_2N(CH_2)_6CH_3$
4. $(C_2H_5)_2N(CH_2)_7CH_3$
5. $(C_2H_5)_2N(CH_2)_8CH_3$

Table II. Effect of diethylaminoalkylbenzenes on carotene content of flavedo of Marsh seedless grapefruit (μg/g dry wt). See caption of Table I for experimental details.

	Control	Compound*				
		1	2	3	4	5
Phytofluene	23.1	25.3	29.3	37.9	38.7	86.2
ζ-Carotene	1.13	1.38	1.38	6.77	27.5	53.8
Neurosporene	0.71	0.83	0.83	1.11	7.16	12.0
Lycopene		8.62	8.62	188	104	153
γ-Carotene		0.55	0.55	1.07	0.59	1.67
α-Carotene	0.57	0.77	0.77	0.61	2.19	0.76
β-Carotene	0.95	0.56	0.56	0.95	trace	0.68
Total	26.5	37.8	37.8	236	180	308
Log _P_		3.07	3.07	4.07	4.57	5.07

*1. $(C_2H_5)_2NCH_2C_6H_5$
2. $(C_2H_5)_2N(CH_2)_2C_6H_5$
3. $(C_2H_5)_2N(CH_2)_3C_6H_5$
4. $(C_2H_5)_2N(CH_2)_4C_6H_5$
5. $(C_2H_5)_2N(CH_2)_5C_6H_5$

Table III. Effect of CPTA and analogs of diethylaminoethylphenylether on carotene content of flavedo of Marsh seedless grapefruit (μg/g dry wt). See caption of Table I for experimental details.

	Control	CPTA	Compound*				
			1	2	3	4	5
Phytofluene	44.9	42.6	48.1	50.9	44.9	51.1	33.7
ζ-Carotene	2.74	10.1	5.45	19.0	13.7	20.6	4.57
Neurosporene	0.33	1.21	0.91	0.84	0.93	1.62	0.63
Lycopene		199	22.2	249	182	226	54.4
γ-Carotene	0.27	1.90	1.01	2.48	2.13	2.74	1.04
α-Carotene	0.35	0.23	0.57	0.41	0.38	0.40	0.63
β-Carotene	1.35	0.39	1.26	0.40	0.82	0.70	0.58
Total	49.9	255	79.5	323	245	303	95.6
Log *P*		4.25	3.05	3.55	4.05	4.35	4.65

*1. $C_6H_5OCH_2CH_2N(C_2H_5)_2$
2. *p*-CH_3-$C_6H_4OCH_2CH_2N(C_2H_5)_2$
3. *p*-C_2H_5-$C_6H_4OCH_2CH_2N(C_2H_5)_2$
4. *p*-*iso*-C_3H_7-$C_6H_4OCH_2CH_2N(C_2H_5)_2$
5. *p*-*tert*-C_4H_9-$C_6H_4OCH_2CH_2N(C_2H_5)_2$

diethylpentylamine and diethylhexylamine, whereas diethylheptylamine caused a larger accumulation. Diethyloctylamine and diethylnonylamine caused very large increases in total carotenoid content. Lycopene accounted for most of the increase in the total carotene content but the intermediates, ζ-carotene (7,8,7',8'-tetrahydro-ψ,ψ-carotene) and neurosporene (7,8-dihydro-ψ,ψ-carotene) also increased appreciably. Of significance are the 3.5 and 4.6-fold increases in the cyclic carotenes, γ-, α- and β-carotene (i.e., β,ψ-, β,ε- and β,β-carotene) caused by diethyloctylamine and diethylnonylamine, respectively. The increase in cyclic carotenes is much larger than that caused by regulators that contained an aromatic ring. Butyldiethylamine has also been observed to cause the development of red color in grapefruit, but only with higher concentrations and longer treatment periods. The higher members of this series, diethyldecylamine, diethylundecylamine, and diethyldodecylamine caused increasing peel injury as the length of the alkyl chain increased. The last two damaged the peel whenever they were applied. The color of the peel next to the damaged areas showed color enhancement, but to a lessening degree as the alkyl chain was lengthened.

Diethylaminoalkylbenzene. The responses of fruit treated with compounds listed in Table II were similar to that of fruit treated with diethylalkylamines (Table I). There was a much larger increase in ζ-carotene and neurosporene for diethylaminobutylbenzene and diethylaminopentylbenzene, whereas the increase in cyclic carotenes was modest. For the diethylalkylamines, lycopene increased to a very high level and then dropped for the last number of the series (Table I). The drop in lycopene accumulation was observed with diethylaminobutylbenzene, but the amount of lycopene increased with diethylaminopentylbenzene, although remaining less than the maximum for this series of bioregulators. The very large increase in phytofluene (15 cis-7,8,11,12,7',8'-hexahydro-ψ,ψ-carotene), as well as ζ-carotene, caused by diethylaminopentylbenzene resulted in a greater increase in the total carotene content. Whether this was caused, in part, by some side effect of the extensive peel damage or entirely by the compound is not certain. A similar effect had been observed previously (14) after treatment with [γ-(diethylamino)-propoxy]-benzene and [δ-(diethylamino)-butoxy]-benzene. In these cases, the increases in phytofluene and ζ-carotene were also very large, but no great peel damage was observed. This effect may arise when the diethylamino and phenyl groups of the inducers are separated by a chain of four or five carbon atoms.

Analogs of Diethylaminoethylphenylethers. Treatment with analogs of diethylaminoethylphenylethers (Table III) gave a similar carotenoid pattern. Neurosporene content did not increase as much as previously, although ζ-carotene content did show a large increase. The cyclic carotenes, with the exception of γ-carotene, did not show a significant increase. There was a general decrease, not previously observed, of all the carotenes after treatment with diethylaminoethyl-4-tert-butylphenylether as compared with diethylaminoethyl-4-isopropylphenylether, instead of the steady increase of ζ-carotene and neurosporene observed in the other series (Tables I and II). The greater biological activity of those compounds in Table III as compared to those in Tables I and II is probably attributable to a higher degree of interaction between the compound and the active site.

Generally, the response increases with increasing concentration of the bioregulator to a maximum value. Further increases in the concentration cause no further increases in carotene content, or the bioregulator can even become somewhat inhibitory to overall carotene synthesis. Doubling of the concentration of diethylaminoethylphenylether from 0.26 to 0.52 M reduced the observed response (14), whereas treatment with CPTA (15) at 0.018 M caused an accumulation of carotenes equal to that in Table III. Doubling the concentration (0.2 M) of diethylaminoethylphenylether and diethylaminoethyl-4-methylphenylether caused large increases of carotenes, whereas doubling of diethylaminoethyl-4-ethylphenylether had almost no effect on carotene content (12). The loss in effectiveness in inducing greater carotene biosynthesis with increasing CPTA concentration has also been observed in *B. trispora* (16).

Inducers of β-Carotenes. A series of *para*-substituted 2-diethylaminoethylbenzoates caused a much larger accumulation of β-carotene than other lycopene inducers, although lycopene remained the major pigment (17). Table IV gives the results of treatment with a series of *para*-substituted 2-diethylaminoethylbenzoates. The response is similar to that previously observed for other triethylamines except for the much larger accumulation of the cyclic carotenes, γ-, α- and particularly β-carotene.

For further explanation of the β-carotene effect, the 2-diethylaminoethylesters of the shorter chain aliphatic acids were investigated (18). As shown in Table V, these esters caused significant increases in β-carotene, so that it was the major pigment, with only a small amount of lycopene. The esters were applied as free amines in isopropanol. The peel remained healthy on all of the fruit except those treated with octanoate and nonanoate which damaged about 30 and 60% of the peel area, respectively. The tendency for peel damage to increase with increasing lipid solubility has been noted for other inducers (19).

As shown in Table VI, the ω-arylaliphatic ester, phenylacetate, more closely resembled the benzoate in its effect than the other members of the series. An increase in the chain length by one methylene group, i.e., hydrocinnamate, reversed the relation between lycopene and β-carotene, and the latter became the major pigment. 4-Phenylbutyrate caused the largest carotenoid accumulation. The most active β-carotene inducers were the hexanoate, 4-phenylbutyrate, and cinnamate.

Table IV. Effect of *para*-substituted 2-diethylaminoethylbenzoates on the carotene content of the flavedo of Marsh seedless grapefruit (μg/g dry wt). Postharvest treatment of fruit is described in reference 17; pigments were isolated and identified by published methods (12). A portion of ground flavedo was dried at 65 C *in vacuo* to obtain the dry wt.

	Control	Compound*									
		1	2	3	4	5	6	7	8	9	10
Phytofluene	173	191	192	229	245	199	244	234	228	233	204
ζ-Carotene	38.6	44.7	60.7	57.6	64.5	44.9	62.5	62.5	60.4	61.1	51.0
Neurosporene	1.52	0.62	0.95	1.23		1.05	1.43	2.01			1.97
Lycopene		1.14	83.6	17.5	215	170	245	285	474	507	
γ-Carotene	0.38	0.37	3.67	1.48	3.82	1.94	3.04	3.80	3.18	2.91	1.14
α-Carotene	0.53	0.53	0.84	1.76	3.86	2.62	2.82	5.54	4.01	4.22	0.40
β-Carotene	1.31	1.51	27.4	12.0	24.7	6.64	13.6	25.0	17.3	32.4	2.05
Total carotenes	215	240	369	320	557	426	572	618	787	841	261
Total xanthophylls	33.7	35.9	39.8	38.7	45.0	44.6	48.5	42.8	45.3	44.6	46.5
Log *P*		1.54	2.75	3.06	3.08	3.14	3.25	3.48	3.93	4.04	4.58

*1. $C_6H_5COOCH_2CH_2N(C_2H_5)_2$
2. *p*-NH_2-$C_6H_4COOCH_2CH_2N(C_2H_5)_2$
3. *p*-CN-$C_6H_4COOCH_2CH_2N(C_2H_5)_2$
4. *p*-NO_2-$C_6H_4COOCH_2CH_2N(C_2H_5)_2$
5. *p*-CH_3O-$C_6H_4COOCH_2CH_2N(C_2H_5)_2$
6. *p*-F-$C_6H_4COOCH_2CH_2N(C_2H_5)_2$
7. *p*-CH_3-$C_6H_4COOCH_2CH_2N(C_2H_5)_2$
8. *p*-Cl-$C_6H_4COOCH_2CH_2N(C_2H_5)_2$
9. *p*-Br-$C_6H_4COOCH_2CH_2N(C_2H_5)_2$
10. *p*-*tert*-C_4H_9-$C_6H_4COOCH_2CH_2N(C_2H_5)_2$

Table V. Effect of 2-diethylaminoethylesters of short chain aliphatic acids on the carotene content of the flavedo of Marsh seedless grapefruit (μg/g dry wt). A portion of ground flavedo was dried at 65 C *in vacuo* to obtain the dry wt. Postharvest treatment of fruits is described in reference 18; pigments were isolated and identified by published methods (12).

	Control	Compound*					
		1	2	3	4	5	6
Phytofluene	36.6	35.9	36.6	32.4	31.2	30.7	39.2
ζ-Carotene	5.11	10.3	19.4	15.7	13.0	8.48	20.5
Neurosporene		0.55		0.61			0.62
Lycopene		9.30	11.4	21.5	8.03	2.79	9.28
γ-Carotene	0.33	1.01	5.57	3.54	2.13	2.01	3.18
α-Carotene	0.33	1.32	3.68	3.05	6.18	3.22	2.22
β-Carotene	1.61	14.2	98.3	58.3	50.5	43.0	50.4
Other carotenes	2.40	1.19	2.52	2.16	2.40	2.04	1.64
Total carotenes	46.4	73.8	177	137	113	92.2	127
Total xanthophylls	24.9	18.2	22.0	24.4	22.1	24.3	21.2

*1. $CH_3(CH_2)_3COOCH_2CH_2N(C_2H_5)_2$
2. $CH_3(CH_2)_4COOCH_2CH_2N(C_2H_5)_2$
3. $CH_3(CH_2)_5COOCH_2CH_2N(C_2H_5)_2$
4. $CH_3(CH_2)_6COOCH_2CH_2N(C_2H_5)_2$
5. $CH_3(CH_2)_7COOCH_2CH_2N(C_2H_5)_2$
6. $C_6H_5CH{=}CHCOOCH_2CH_2N(C_2H_5)_2$

Table VI. Effect of aryl and arylaliphatic esters of 2-diethylaminoethanol on the carotene content of the flavedo of Marsh seedless grapefruit (μg/g dry wt). See caption of Table V for experimental details.

	Control	Compound*			
		1	2	3	4
Phytofluene	47.8	43.9	47.9	35.9	34.7
ζ-Carotene	6.48	11.0	10.7	9.91	9.78
Neurosporene	0.41	1.07	0.84	1.20	1.37
Lycopene		31.6	22.9	4.72	13.1
γ-Carotene	0.34	1.10	0.86	1.04	3.90
α-Carotene	0.18	0.38	0.81	0.78	5.48
β-Carotene	0.81	1.97	3.85	23.6	93.4
Other carotenes	1.68	0.81	0.52	0.34	0.72
Total carotenes	57.7	91.9	88.4	77.5	162
Total xanthophylls	29.8	33.7	33.7	26.6	29.0

*1. $(C_2H_5)_2NCH_2CH_2OOCC_6H_5$
2. $(C_2H_5)_2NCH_2CH_2OOCCH_2C_6H_5$
3. $(C_2H_5)_2NCH_2CH_2OOCCH_2CH_2C_6H_5$
4. $(C_2H_5)_2NCH_2CH_2OOCCH_2CH_2CH_2C_6H_5$

Essential Structural Features for Induction of trans-Carotenes. As a further elaboration on the general formula $(C_2H_5)_2NCH_2R$, given for trans-lycopene-inducing bioregulators, none of the following structural alterations totally abolishes the activity or changes the response pattern: (1) replacement of the S-atom by oxygen which connects the amine portion and the benzene moiety in the CPTA molecule; (2) elimination of the substitution on the benzene ring; (3) elimination of the benzene ring itself; (4) replacement of the ethyl groups of the amine portion by other alkyl groups. However, these alterations changed the effectiveness of the modified bioregulators. Benzene derivatives were more effective than alkyl derivatives and the substitutions of certain groups on the benzene ring at the para position seem to make the compounds more effective.

As the data in Table VII show, substitution of a methyl group at the ortho position almost completely eliminated the carotenoid-inducing ability of the compound. Substitution at the meta position also causes a reduction in the inducing ability. On the other hand, with a group in the para position, the compound was a very powerful inducer. The same effect was observed with the strongly electron-withdrawing chloro group, as compared to the electron releasing methyl groups, for p-, m-, and o-chlorophenoxy triethylamine (19). Steric hindrance may possibly play an important role in the effectiveness of the position isomers. Studies have indicated that the effectiveness of the compounds does not depend solely upon the electron-withdrawing ability of the substituting groups on the benzene moiety (20). Also, experimental data obtained thus far do not necessarily rule out the possible induction effects of substituents on the amine portion of the molecule.

It should be emphasized that the variable effectiveness on carotenogenesis among the triethylamine derivatives suggests that certain compounds did not penetrate well into the fruit flavedo tissue. Thus, the lipophilic/hydrophilic properties of the bioregulators appear to influence the effectiveness of the compounds. These characteristics are reflected in the value of log P, which is the logarithm of the octanol/water partition coefficient of the unionized molecule (21, 22, 23). Log P reflects the ability of the compound to pass through the various aqueous and lipid layers in the cell and, therefore, should correlate with the concentration of bioregulator at the regulatory site. Log P values (Tables I-IV) have proven useful, in combination with considerations of the electronic and steric states of a molecule, as a general guide in the designing of new bioregulators (12, 19). Thus, there is an upper limit for log P at which peel damage begins to occur. The compounds with log P greater than 4.6 probably cause peel damage by disrupting the lipid membranes of the cells. Those compounds with log P less than 4.6 show no damage. Certain of the compounds probably interact more strongly at the active site(s) and produce a noticeable

Table VII. Effect of steric position of the methyl group in diethylaminoethylmethylphenylether on the carotene content of the flavedo of Marsh seedless grapefruit (μg/g dry wt). A portion of ground flavedo was dried at 65 C *in vacuo* to obtain the dry wt. Postharvest treatment of fruits is described in reference 19; pigments were isolated and identified by published methods (12).

	Control	Compound		
		o-CH_3	m-CH_3	p-CH_3
Phytofluene	40.2	39.3	36.1	52.7
ζ-Carotene	4.66	3.26	4.00	17.7
Neurosporene	0.26	0.79	1.14	1.77
Lycopene		0.43	24.4	516
γ-Carotene		0.22	0.60	0.83
α-Carotene	0.52	0.34	0.61	0.28
β-Carotene	1.25	0.86	0.59	0.48
Total carotenes	46.9	45.2	67.4	589
Total xanthophylls	21.8	21.2	24.3	25.1

effect on carotenogenesis at lower concentrations and for smaller values of log *P* than others. The optimum value for log *P* appears to be in the range of 3.5 to 4.5.

Mode of Action of trans-Carotenoid Bioregulators. The fact that carotenogenic microorganisms, such as *B. trispora* and *P. blakesleeanus*, respond to treatment (14) provided an opportunity to study the mechanism of action of bioregulators in carotenoid formation. Studies (16) conducted using the mold *B. trispora* indicated that the bioregulators act at the enzyme level by inhibiting cyclase(s); transformation of the acyclic lycopene to the monocyclic γ-carotene and the bicyclic β-carotene is inhibited. These studies also suggested that the bioregulators induce (derepress) a gene that regulates the synthesis of a specific enzyme, or enzymes, in the primary biosynthetic pathway of carotenoids and thus increase net synthesis of carotenoid pigments. The nullifying action of cycloheximide on the effect of the bioregulatory agent in carotenoid synthesis indicated that the latter compounds act as an inducer (derepressor) of enzyme synthesis rather than as an activator of pre-existing enzyme(s). Once the enzymes that participate in carotenoid synthesis have been formed, cycloheximide does not affect their activity. Cycloheximide is known to act at the ribosomal level to inhibit protein synthesis (24). The relative effectiveness of an individual bioregulator as a cyclase inhibitor and as an enzyme inducer could lead to variations in the carotenoid pigment response pattern among the different bioregulators, as seen in Tables I-VII.

cis-Carotenes

All of the compounds reported above cause the accumulation of the normal, all-*trans* carotenoids. They are of the general formula $(C_2H_5)_2NCH_2R$. The carotene content greatly increases and all-*trans* lycopene becomes the major pigment unless the tertiary amines are aliphatic esters of 2-diethylaminoethanol. If they are, all-*trans* β-carotene is the predominant carotene.

Studies of structure-activity relationships of compounds that affect carotenogenesis led to the discovery of a new class of bioregulators that cause the accumulation of poly-*cis*-carotenoids (25, 26). The natural occurrence of *cis* and poly-*cis*-carotenoids has been observed, but is not very common (27-32).

Tables VIII and IX show that all of the substituted dibenzylamines, *N*-benzylphenethylamine, and *N*-benzyl-2-naphthalenemethylamine, and all of the secondary amines stimulated the production of poly-*cis*-carotenes. Tables X, XI, and XII show the ability of substituted *N*-benzylfurfurylamines; *N*-benzyl,*N*-methylfurfurylamines; and *N*-alkyl,*N*-methylbenzylamines to stimulate the formation of poly-*cis*-carotenes.

The pattern of accumulation of the poly-*cis*-carotenes is the same as that caused by the benzylamines. The *N*-benzylfurfurylamines are more effective than the corresponding dibenzylamines. The only anomaly was 2-(4-bromophenoxy)-

Table VIII. Effect of dibenzylamine and substituted dibenzylamines on the carotene content of the flavedo of Marsh seedless grapefruit (μg/g dry wt). A portion of ground flavedo was dried at 65 C *in vacuo* to obtain the dry wt. Postharvest treatment of fruits is described in reference *25*; pigments were isolated and identified by published methods (*25*).

	Control	Compound*						
		1	2	3	4	5	6	7
Phytofluene	35.0	36.1	39.2	40.1	37.9	32.2	50.3	43.3
α-Carotene	0.14	0.11	0.10	0.16	0.12	0.11	0.14	0.09
β-Carotene	0.65	0.83	0.95	1.88	1.05	1.00	1.01	0.78
ζ-Carotene	4.37	10.12	12.64	21.2	18.6	14.0	18.7	12.5
Poly-*cis*-γ-carotene I†	0.31	0.70	1.35	3.18	3.84	2.50	1.49	0.78
Proneurosporene†		4.91	7.99	16.2	15.0	9.30	13.5	8.28
Prolycopene†	0.55	8.03	14.1	33.1	29.8	21.3	25.0	13.2
cis-Lycopene†	1.21	5.12	7.58	9.69	9.31	9.45	10.1	6.00
Poly-*cis*-γ-carotene II†	0.87	3.35	5.48	11.9	6.81	5.55	5.91	4.50
Unknown 453	0.48	1.74	1.44	4.18	8.22	4.11	6.40	0.94
Total carotenes	43.6	70.0	90.8	41.7	30.7	99.5	132	90.3
Total xanthophylls	25.5	28.8	30.0	39.4	36.6	26.1	39.7	28.5

*1. $C_6H_5CH_2NHCH_2C_6H_5$
2. $4\text{-}F\text{-}C_6H_4CH_2NHCH_2C_6H_5$
3. $4\text{-}Cl\text{-}C_6H_4CH_2NHCH_2C_6H_5$
4. $4\text{-}Br\text{-}C_6H_4CH_2NHCH_2C_6H_5$
5. $4\text{-}CH_3\text{-}C_6H_4CH_2NHCH_2C_6H_5$
6. $4\text{-}NO_2\text{-}C_6H_4CH_2NHCH_2C_6H_5$
7. $4\text{-}CN\text{-}C_6H_4CH_2NHCH_2C_6H_5$

†Tentative assignment of poly-*cis* carotenes in control.

Table IX. Effect of substituted dibenzylamines on the carotene content of the flavedo of Marsh seedless grapefruit (μg/g dry wt). See caption of Table VIII for experimental details.

	Control	Compound*				
		1	2	3	4	5
Phytofluene	38.8	48.8	41.5	40.7	33.1	43.1
α-Carotene	0.11	0.14	0.18	0.13	0.09	0.17
β-Carotene	0.62	0.84	1.07	1.44	1.08	1.08
ζ-Carotene	5.20	12.0	10.75	18.3	10.2	11.1
Poly-<u>cis</u>-γ-carotene I†	0.54	0.71	1.20	3.32	1.49	1.56
Proneurosporene†		6.55	5.44	11.7	5.13	5.51
Prolycopene†	1.31	7.72	8.90	25.23	9.86	10.5
<u>cis</u>-Lycopene	1.45	5.57	5.57	14.0	7.68	7.75
Poly-<u>cis</u>-γ-carotene II†	0.80	2.45	4.12	8.37	3.67	2.42
Unknown 453		1.54	1.18	3.29	2.37	2.98
Total carotenes	43.8	86.3	79.9	126	74.6	86.2
Total xanthophylls	24.6	34.8	35.5	40.0	31.9	37.2

*1. $2\text{-}CH_3\text{-}C_6H_4CH_2NHCH_2C_6H_5$
2. $3\text{-}CH_3\text{-}C_6H_4CH_2NHCH_2C_6H_5$
3. $4\text{-}CH_3\text{-}C_6H_4CH_2NHCH_2C_6H_5$
4. $(4)Cl\text{-}C_6H_4CH_2NHCH_2C_6H_4\text{-}CH_3(4')$
5. $(4)CH_3\text{-}C_6H_4CH_2NHCH_2C_6H_4\text{-}NO_2(4')$

†Tentative assignments of poly-<u>cis</u> carotenes in control.

Table X. Effect of furfurylamine analogs on the carotene content of the flavedo of Marsh seedless grapefruit (μg/g dry wt). A portion of ground flavedo was dried at 65 C *in vacuo* to obtain the dry wt. Solutions of test compounds were prepared at 0.1 M in isopropanol and were applied as the free amines. The postharvest treatment is as described in reference 12. The flavedo was removed 14 days after initial treatment; pigments were isolated and identified by published methods (25).

	Control	Compound*					
		1	2	3	4	5	6
Phytofluene	19.8	22.0	21.2	25.0	31.0	19.4	25.6
α-Carotene	0.34	0.24	0.33	0.23	0.20	0.21	0.22
β-Carotene	1.54	1.26	1.36	1.31	1.46	0.93	1.48
ζ-Carotene	2.58	11.7	6.18	18.7	27.6	2.26	22.8
Poly-*cis*-γ-carotene I		0.95	0.45	2.24	2.72		2.01
Proneurosporene		8.41	3.42	15.2	21.4	0.44	18.3
Prolycopene		11.4	5.23	32.9	46.5	0.72	41.5
γ-Carotene							
cis-Lycopene		7.31	3.81	14.8	23.4	1.13	17.5
Neurosporene	1.17					0.72	
Poly-*cis*-γ-carotene II		4.82	1.68	10.5	14.5		8.46
Unknown 453		2.03	1.16	2.05	3.70		3.54
Lycopene							
Total carotenes	25.4	70.2	44.8	123	173	25.8	142
Total xanthophylls	19.1	22.2	22.4	24.8	30.1	20.6	27.1

*1. $C_6H_5CH_2NHCH_2$-R
2. 4-CH_3-$C_6H_4CH_2NHCH_2$-R
3. 4-Cl-$C_6H_4CH_2NHCH_2$-R
4. 4-Br-$C_6H_4CH_2NHCH_2$-R
5. 4-NO_2-$C_6H_4CH_2NHCH_2$-R
6. $C_6H_5CH_2CH_2NHCH_2$-R

(R = 2-furyl ring structure)

Table XI. Effect of analogs of benzylfurfurylamine and methylfurfurylamine on the carotene content of the flavedo of Marsh seedless grapefruit (μg/g dry wt). See caption of Table X for experimental details.

	Control	Compound*					
		1	2	3	4	5	6
Phytofluene	22.1	28.8	31.5	35.6	33.3	30.9	37.1
α-Carotene	0.13	0.19	0.20	0.23	0.30	0.14	0.17
β-Carotene	0.57	1.11	1.46	1.06	1.28	1.50	1.73
ζ-Carotene	2.68	20.7	29.8	21.1	21.5	37.0	35.8
Poly-cis-γ-carotene I		2.82	4.05	4.47	3.08	3.01	5.55
Proneurosporene		14.7	21.4	19.3	14.5	30.0	25.5
Prolycopene		27.6	48.2	45.0	33.3	79.5	65.3
cis-Lycopene		18.1	29.0	26.8	26.2	31.0	39.2
Poly-cis-γ-carotene II		13.6	21.0	21.7	10.1	18.1	21.4
Unknown 453		7.12	8.55	5.67	8.10	5.92	9.09
Total carotenes	25.5	135	195	179	152	231	241
Total xanthophylls	26.3	35.7	42.0	36.7	38.9	32.1	46.8

*1. $C_6H_5CH_2NHCH_2-R$
2. $C_6H_5CH_2N<^{CH_3}_{CH_2-R}$
3. $C_6H_5CH_2NHCH_2-R$
4. $4-Br-C_6H_4CH_2N<^{CH_3}_{CH_2-R}$
5. $C_6H_5CH_2CH_2NHCH_2-R$
6. $C_6H_5CH_2CH_2N<^{CH_3}_{CH_2-R}$

(R = 2-furyl [furan ring structure])

Table XII. Effect of N-alkyl,N-methylbenzylamines on the carotene content of the flavedo of Marsh seedless grapefruit (μg/g dry wt). See caption of Table X for experimental details.

	Control	Compound*				
		1	2	3	4	5
Phytofluene	24.9	29.4	24.0	35.2	28.1	34.1
α-Carotene	0.54	0.36	0.23	0.23	0.21	0.31
β-Carotene	0.95	1.01	1.70	1.64	0.77	0.93
ζ-Carotene	2.73	3.85	7.26	21.2	20.9	25.0
Poly-cis-γ-carotene I			0.86	2.94	2.90	3.15
Proneurosporene		1.77	5.01	16.1	16.4	18.8
Prolycopene		2.66	9.45	32.6	36.1	26.4
cis-Lycopene		3.63	4.75	11.0	13.3	12.3
Poly-cis-γ-carotene II		0.39	3.58	11.2	8.03	9.83
Unknown 453		0.20	1.21	2.09	3.58	2.03
Total carotenes	29.0	43.3	58.1	134	130	133
Total xanthophylls	25.8	25.0	27.9	33.9	27.9	27.2

*1. $C_6H_5CH_2N<^{CH_3}_{CH_2(CH_2)_2CH_3}$

2. $C_6H_5CH_2N<^{CH_3}_{CH_2(CH_2)_3CH_3}$

3. $C_6H_5CH_2N<^{CH_3}_{CH_2(CH_2)_4CH_3}$

4. $C_6H_5CH_2N<^{CH_3}_{CH_2(CH_2)_5CH_3}$

5. $C_6H_5CH_2N<^{CH_3}_{CH_2(CH_2)_6CH_3}$

ethylfurfurylamine which acted as an inducer of both poly-*cis*-carotenes and lycopene. It also inhibited the cyclase(s), a characteristic of lycopene inducers, and prevented the formation of poly-*cis*-γ-carotene I and II. The effectiveness of the compounds *N*-methyl,*N*-benzylfurfurylamine; *N*-methyl-,*N*-4-bromobenzylfurfurylamine; and *N*-methyl,*N*-benzylfurfurylamine shows that *N*-methylation of the corresponding secondary amines did not eliminate, but tended to enhance activity. This enhancement of activity is especially evident with the *N*-alkyl,*N*-methylbenzylamines. Whereas *N*-(*n*-hexyl)benzylamine is only weakly active, *N*-methyl,*N*-(*n*-hexyl)benzylamine was a very good inducer. These compounds are not good inhibitors of the cyclase(s) as is the case with the lycopene inducers. In none of the samples was any trace of all-*trans* lycopene found. Prolycopene was the major pigment, but there were relatively large amounts of other poly-*cis*-carotenes. In marked contrast, the triethylamine bioregulators primarily induced the synthesis of all-*trans* lycopene or β-carotene. Although not experimentally determined, the increase in the total xanthophylls observed in the treated fruit could have been caused by the formation of *cis* and poly-*cis*-xanthophylls. *cis*-Xanthophylls do occur naturally in some plants (30).

Mode of Action of cis-Carotenoid Bioregulators. The mode of action of the poly-*cis* inducers is probably gene derepression, similar to that of the lycopene inducers. However, these compounds probably derepress a recessive gene that controls the biosynthesis of the poly-*cis*-carotenoids, whereas the lycopene inducers derepress the dominant genes giving rise to the normal all-*trans* carotenoids. The existence of this recessive gene can only be postulated as present in grapefruit and other citrus. Important genetic studies have shown that such a gene exists in the tomato (*Lycopersicum esculentum*) (28). Tangerine tomatoes, which are homozygous for the recessive allele t, accumulate the orange pigment prolycopene at the expense of lycopene. Red fruit carry the dominant allele t+. The same situation could exist in other fruit and explain the action of the poly-*cis* inducers. This interpretation lends support to the postulated parallel pathways for all-*trans* and poly-*cis*-carotenoids (30). If this hypothesis is true, one would also expect to find poly-*cis*-ζ-carotene, -phytofluene, and -phytoene present in the fruit. Poly-*cis* induction appears to differ from the lycopene inducer in another respect. In addition to gene derepression, the lycopene inducers inhibit the cyclase(s), causing lycopene to accumulate at the expense of the cyclic carotenes. The accumulation of significant amounts of poly-*cis*-γ-carotenes I and II indicates that the cyclase(s) is not inhibited by the poly-*cis* inducers and that their only apparent function is to derepress the recessive gene.

Literature Cited

1. Mackinney, G.; Nakayama, T.; Buss, C. D.; Chichester, C. O. J. Am. Chem. Soc. 1952, 74, 3156.
2. Mackinney, G.; Chichester, C. O.; Wong, P. S. J. Am. Chem. Soc. 1953, 75, 5428.
3. Mackinney, G.; Nakayama, T.; Chichester, C. O.; Buss, C. D. J. Am. Chem. Soc. 1953, 75, 236.
4. Engle, B. G.; Wursch, J.; Zimmerman, M. Halv. Chem. Acta 1953, 36, 1771.
5. Chichester, C. O.; Wong, P. S.; Mackinney, G. Plant Physiol. 1954, 29, 238.
6. Caglioti, L.; Cainelli, G.; Camerino, G.; Mondelli, R.; Prieto, A.; Quilico, A.; Salvatori, T.; Selva, A. Chim. Ind. (Milan) 1964, 46, 1.
7. Thomas, D. M.; Goodwin, T. W. Phytochemistry 1967, 6, 355.
8. Yokoyama, H.; Hsu, W. J. Unpublished data.
9. Ninet, L.; Renauet, J.; Tissier, R. Biotech. Bioeng. 1969, 11, 2985.
10. Knypl, J. S. Naturwissenschaften 1969, 56, 572.
11. Coggins, C.; Henning, G. L.; Yokoyama, H. Science 1970, 168, 1589.
12. Poling, S.; Hsu, W. J.; Yokoyama, H. Phytochemistry 1975, 14, 1933.
13. Yokoyama, H.; White, M. J. J. Agr. Food Chem. 1967, 15, 693.
14. Poling, S.; Hsu, W. J.; Yokoyama, H. Phytochemistry 1973, 2, 2665.
15. Yokoyama, H.; DeBenedict, C.; Coggins, C. W.; Henning, G. L. Phytochemistry 1972, 11, 1721.
16. Hsu, W. J.; Yokoyama, H.; Coggins, C. W. Phytochemistry 1972, 11, 2895.
17. Poling, S. M.; Hsu, W. J.; Yokoyoma, H. Phytochemistry 1976, 15, 1685.
18. Poling, S. M.; Hsu, W. J.; Koehrn, F. J.; Yokoyama, H. Phytochemistry 1977, 16, 551.
19. Hsu, W. J.; Poling, S. M.; DeBenedict, C.; Rudash, C.; Yokoyama, H. J. Agric. Food Chem. 1975, 23, 831.
20. Hansch, C.; Muir, R. M.; Fujita, T.; Maloney, P. P.; Geiger, F.; Streich, M. J. Am. Chem. Soc. 1963, 85, 2817.
21. Hansch, C.; Fujita, T. J. Am. Chem. Soc. 1964, 86, 1616.
22. Fujita, T.; Iwasa, J.; Hansch, C. J. Am. Chem. Soc. 1964, 86, 5175.
23. Hansch, T.; Quinlan, J. E.; Lawrence, G. L. J. Org. Chem. 1967, 33, 347.
24. Boulter, D. Ann. Rev. Plant Physiol. 1970, 21, 91.
25. Poling, S. M.; Hsu, W. J.; Yokoyama, H. Phytochemistry 1980, 19, 1677.
26. Poling, S. M.; Hsu, W. J.; Yokoyama, H. Phytochemistry (In press).

27. Zechmeister, L. "cis-trans Isomeric Carotenoids, Vitamins A, and Arylpolyenes"; Academic Press: New York, 1962; 251 p.
28. Goodwin, T. W. "Carotenoids"; Isler, O., Ed.; Birkhauser-Verlag: Basel, 1971; p. 577.
29. Weedon, B. C. L. "Carotenoids"; Isler, O., Ed.; Birkhauser-Verlag: Basel, 1971; p. 267.
30. Raymundo, L. C.; Simpson, K. L. Phytochemistry 1972, 11, 397.
31. Glass, R. W.; Simpson, K. L. Phytochemistry 1976, 15, 1077.
32. Zechmeister, L.; Schroeder, W. A. J. Am. Chem. Soc. 1942, 64, 1173.

RECEIVED September 22, 1981.

10

Biochemical Effects of Glyphosate [*N*-(Phosphonomethyl)glycine]

ROBERT E. HOAGLAND and STEPHEN O. DUKE

Southern Weed Science Laboratory, Stoneville, MS 38776

Glyphosate [N-(phosphonomethyl)glycine] is implicated in the biochemical alteration of various processes in plants and microorganisms, however, its inhibition of aromatic amino acid biosynthesis is the only well established primary mode of action of this herbicide. The enzyme 5-enolpyruvylshikimate-3-phosphate synthase is inhibited by physiological concentrations of glyphosate and is the most sensitive site of action of glyphosate in reducing aromatic amino acid levels. Aromatic amino acid depletion reduces or stops protein synthesis, causing cessation of growth and eventually cellular disruption and death. Supplemental aromatic amino acids reverse glyphosate-caused growth inhibition in microorganisms and in *Lemna*, however, they are not always antidotal, especially with intact terrestrial plants. In higher plants, glyphosate increases extractable phenylalanine ammonia-lyase (PAL) activity which partially explains observed reduced phenylalanine and tyrosine pools. Still, PAL inhibitor studies and feeding experiments indicate that the mode of action of glyphosate in intact higher plants cannot be solely explained by interference with phenolic metabolism. Glyphosate's divalent metal cation chelation properties may also be important in many biochemical interactions. Glyphosate also disrupts chloroplasts, membranes, and cell walls; alters protein and nucleic acid synthesis, photosynthesis, and respiration; and reduces chlorophyll, and other porphyrin compound synthesis. Whether these effects are primary or secondary is not yet established. The rapid biochemical effects of glyphosate on many parameters, and its highly non-specific toxicity indicate that this herbicide may have multiple primary action sites.

General aspects of glyphosate [N-(phosphonomethyl)glycine], including chemical and herbicidal properties are important to consider, because they assist in understanding glyphosate's biochemical effects. Absorption, translocation, and degradation of glyphosate and the effects of glyphosate on growth and the associated phytotoxic symptoms will be briefly presented, followed by an in-depth discussion of the biochemical action of glyphosate. An attempt to point out what we consider to be primary actions and secondary effects will be made in the summary section.

Discovery and Development of Glyphosate

Glyphosate has developed into an extremely important herbicide since its introduction in 1971 (1). It has a simple molecular structure, a relatively high water solubility, and a low molecular weight compared to most herbicides (Figure 1; Table I). Roundup® (Monsanto's herbicide formulation of the isopropylamine salt of glyphosate with a surfactant) is now used extensively in various crop and non-agricultural situations in many countries of the world. Glyphosate is a non-selective, broad spectrum, postemergence herbicide and is the only compound of this chemical class which is registered as a herbicide; however, an analog, glyphosine (Figure 2) is a plant growth regulator.

About 1200 literature citations (including published articles and abstracts) exist on various aspects of glyphosate, but there are few review articles on the compound. A description of the compound and its general herbicidal properties was published soon after its introduction (1). Updates of the compound's selectivity and characteristics have since been published (2, 3, 4). Franz (5) covered both general and specific aspects of glyphosate, including the relationship between structure and activity of glyphosate derivatives and related compounds as well as various proposed modes of action. A bibliography of glyphosate literature to 1978 is available (6). The Herbicide Handbook of the Weed Science Society of America outlines glyphosate's uses, properties, precautions, physiological and biochemical behavior, synthesis, analytical methods and procedures, etc. (7). A very brief general overview of glyphosate including biochemical and physiological action was published recently (8). A comprehensive review of glyphosate is presently in preparation (9). A chapter that presents

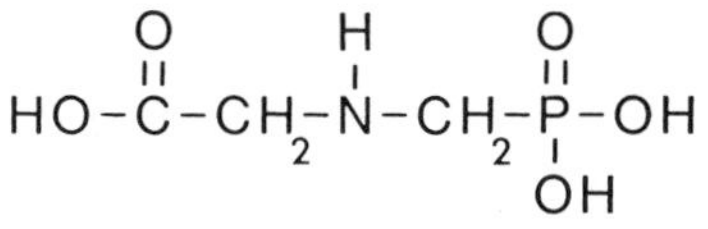

$$HO-\overset{\overset{O}{\|}}{C}-CH_2-\overset{\overset{H}{|}}{N}-CH_2-\underset{\underset{OH}{|}}{\overset{\overset{O}{\|}}{P}}-OH$$

N-(Phosphonomethyl)glycine

Figure 1. Glyphosate chemical structure.

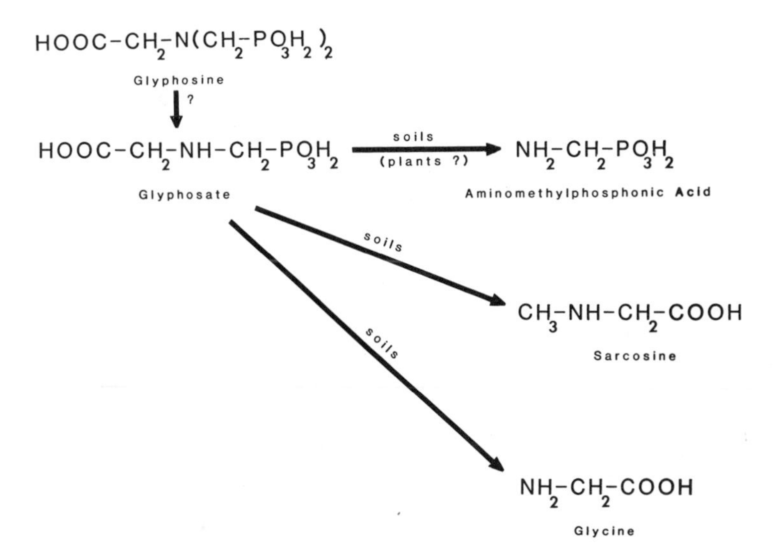

Figure 2. Analogs, metabolites, and/or degradation products of glyphosate.

Table I. General characteristics and properties of glyphosate.

		PHYSICAL AND CHEMICAL PROPERTIES
Physical state	-	white odorless solid
Molecular weight	-	169.1
Solubility	-	H_2O (1 to 8%, 25-100°C)
Melting point	-	200°C w. decomposition
Vapor pressure	-	negligible
Density	-	0.5 g/cc
pK_a 1, 2, 3	-	2.3, 5.9, 10.9
Chelation	-	metal cations
		STABILITY ASPECTS
Photodecomposition	-	negligible
Shelf life	-	very stable
Biodegradability and soil persistence	-	rapidly degraded by soil microoganisms $T_{1/2}$ < 60 da.
		UPTAKE AND METABOLIC ASPECTS
Absorption	-	readily absorbed by roots and foliage
Translocation	-	rapidly translocated from application point
Metabolism	-	metabolism by plants is extremely low, but rapid in soils
Specificity	-	non-selective, broad spectrum
		TOXICITY
Rats	-	LD_{50} = 4320 mg/kg
Rabbits	-	MLD = > 7940 mg/kg
Fish	-	EC_{50} = > 1000 ppm

various aspects of glyphosate has been recently included in a text on herbicide mode of action (10).

The overall success of glyphosate may be attributed to several properties (Table I) in addition to its phytotoxicity. Its low molecular weight and high water solubility are factors that aid in its rapid absorption and translocation by plant tissues. Glyphosate is a white, odorless solid, and as a phosphonic acid has the ability to chelate certain divalent and trivalent cations (11-14). Glyphosate is rapidly absorbed by foliar tissues and roots and can be relatively rapidly translocated to various plant organs, distant from the application site (15-18). Once inside the plant, glyphosate does not break down nor is it metabolized to a significant degree (16, 19-21). Thus, it can maintain its phytotoxic action while translocating to various plant parts. These factors provide utility in controlling hard-to-kill perennial weeds that are deep-rooted or ones that possess vegetative propagules.

In soils, however, the compound is strongly adsorbed, and is rapidly degraded by microorganisms to non-phytotoxic products including carbon dioxide (7, 17, 22). Glyphosate is very stable and is not subject to photodecomposition or volatility. Due to its biodegradability, it has a half-life of less than 60 days in soil. Although the compound is a highly efficacious herbicide, its toxicological effects on mammals, honeybees, and fish, are relatively low (5, 7) and it has little effect on diatom and *Daphnia* populations in aquatic environments (23, 24).

Non-biochemical Considerations

Phytotoxic Symptoms. In diverse species, the first visible growth effect of glyphosate after application is generally the induction of chlorosis, which is usually followed by necrosis (15, 16, 25-33). These symptoms commonly take from two to ten days to occur and in some perennials, injury symptoms may be evident in the year following treatment. Morphological abnormalities of leaves and wilting have been observed under some conditions (16, 27, 30, 34). Root and rhizome growth and survival are strongly inhibited by glyphosate (18, 25, 27, 35-42). No particular plant organ or tissue has been conclusively shown to be the primary site of action of glyphosate, however, the effects on meristems are perhaps directly caused by the herbicide because they have high metabolic activity and glyphosate accumulates there (40).

Chlorosis is often induced more rapidly under high than under low light intensity (43-46). Morphological differences between sun and shade leaves, however, can result in greater toxicity to shaded than to unshaded plants sprayed with glyphosate (35). Increased light intensity has also been correlated with increased glyphosate accumulation in untreated plant tissue parts (47).

Although seedling growth is inhibited by glyphosate, this herbicide has no significant effect on germination of a wide variety of species (40, 44, 48-51). Because glyphosate is tightly bound to soil (52) and is also rapidly metabolized in soil, its availability to germinating seeds may be minimized.

Absorption and Translocation. Absorption and translocation of glyphosate is summarized and discussed in depth elsewhere (10). Glyphosate uptake and translocation is relatively rapid in diverse species (15-19, 21, 53, 54). Environmental factors such as temperature and relative humidity have been studied with respect to glyphosate phytotoxicity, uptake, and translocation (20, 21, 35, 36, 43, 55-58). Growth stage and water stress effects on glyphosate's mobility in bermudagrass [*Cynodon dactylon* (L.) Pers.] has also been investigated (59). Although such factors can alter the rates of absorption and translocation of glyphosate, the herbicide is generally rapidly absorbed and translocated to various plant tissues. Meristems are known sites of glyphosate accumulation (40). Translocation to underground propagules of perennial species prevents regrowth from these sites and results in their subsequent destruction (19, 20). Information available from absorption and translocation studies suggests that most glyphosate movement is in the symplast but there is also some evidence of apoplastic transport (15, 17, 18, 21, 38).

Degradation of Glyphosate. Glyphosate is very stable in higher plants (16, 19-21, 53), but is degraded by microorganisms in soil (7, 17, 22). Various metabolites or degradation products of glyphosate have been identified, tentatively identified, or proposed (Figure 2). Aminomethylphosphonic acid is the principle product of glyphosate degradation in soils (60). This compound has been found in plants, but was absorbed from the soil and did not result from metabolic action on glyphosate in the plant. Sarcosine and glycine are other possible non-phytotoxic products of glyphosate degradation in soils (60). Radiolabeled glyphosate has been shown to degrade completely in the soil to carbon dioxide (17, 22).

Biochemical and Physiological Effects

Early Work and Feeding Studies. Although many biochemical and physiological investigations have been conducted on glyphosate effects and action in plants, the most promising of these have implicated a disruption of phenolic metabolism as the basis for its molecular mode of action. An outline of these compounds and associated enzymes of phenolic metabolism in higher plants is presented in Figure 3. Enzymes 1, 2, 3,

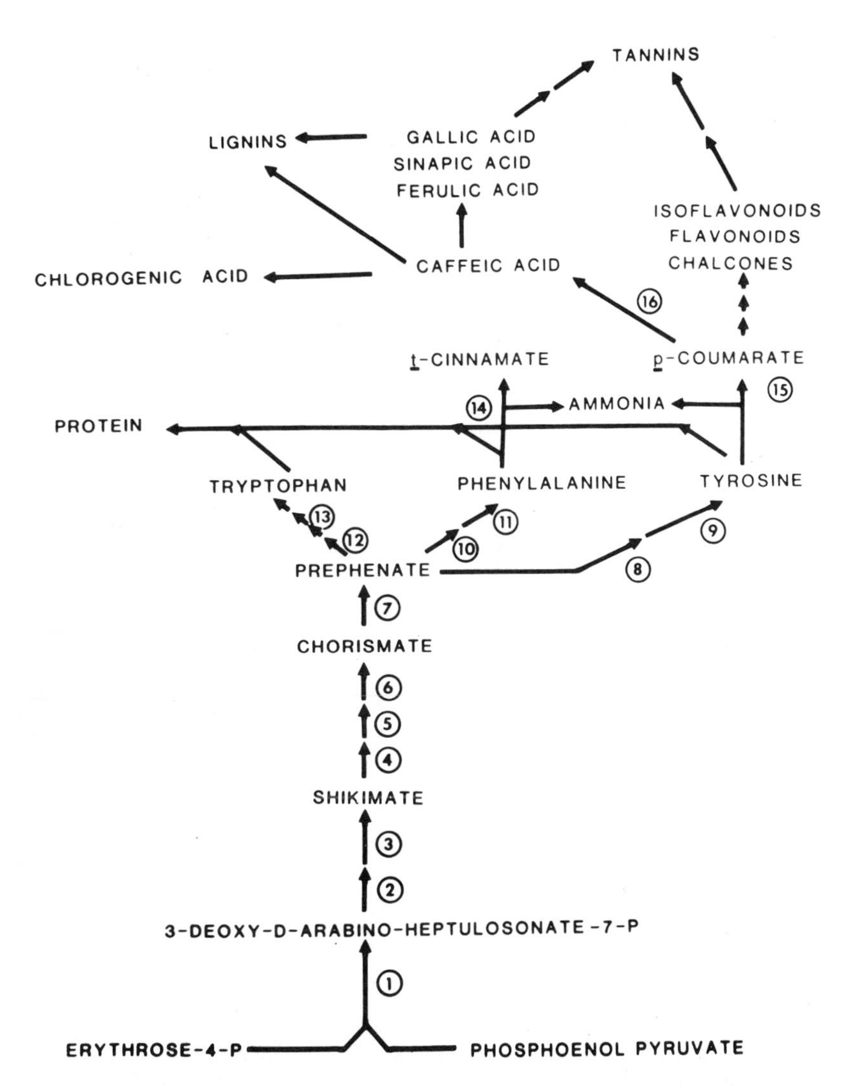

Figure 3. Schematic outline of various intermediates and products including enzymes of the phenolic pathway in plants. Enzymes: 1, 3-deoxy-2-oxo-D-arabinoheptulosate-7-phosphate synthase; 2, 5-dehydroquinate synthase; 3, shikimate dehydrogenase; 4, shikimate kinase; 5, 5-enolpyruvylshikimate-3-phosphate synthase; 6, chorismate synthase; 7, chorismate mutase; 8, prephenate dehydrogenase; 9, tyrosine aminotransferase; 10, prephenate dehydratase; 11, phenylalanine aminotransferase; 12, anthranilate synthase; 13, tryptophan synthase; 14, phenylalanine ammonia-lyase; 15, tyrosine ammonia-lyase; and 16, polyphenol oxidase.

5, 7, 8, 10, 14, 15, and 16 have received the most attention with respect to glyphosate interaction (Table II).

In early investigations of glyphosate on the aquatic plant, duckweed (*Lemna gibba* L.), and the microorganism, *Rhizobium japonicum*, Jaworski indicated that the herbicide caused decreased levels of aromatic amino acids (61). Feeding supplemental aromatic amino acids resulted in reversal of glyphosate-caused inhibition of growth. These studies indicated that glyphosate caused deficits of phenylalanine, tyrosine, and to a lesser degree, tryptophan. The theory was thus proposed that glyphosate inhibited the synthesis of these amino acids by inhibiting or repressing the enzymes chorismate mutase (Figure 3, No. 7) and prephenate dehydratase (Figure 3, No. 10) in *Lemna gibba* (Figure 4). In addition to these two enzymes, prephenate dehydrogenase (Figure 3, No. 8) was also proposed to be inhibited in *Rhizobium japonicum* (Figure 4). Depletion of aromatic amino acids pools could lead to reduced protein synthesis, resulting in cessation of growth, cellular disruption, and eventually death. This original theory has been strongly supported in most cases by subsequent experiments with other microorganisms, plant cell tissue cultures, and isolated plant tissue systems. Partial or complete reversal of glyphosate-caused growth inhibition by aromatic amino acids has been shown in the unicellular organisms *E. coli* (62, 63), *Rhizobium japonicum* (61), and *Chlamydomonas* (63); and in tissue cultures of carrot (*Daucus carota* L.) (63, 64), soybeans [*Glycine max* (L.) Merr.] (63), and tobacco (*Nicotiana tabacum* L.) (65). In isolated soybean leaf cells, supplemental addition of aromatic amino acids partially prevented glyphosate-reduced protein synthesis (66). Feeding aromatic amino acids has also reversed glyphosate-induced basal-stem swelling and bud release in grain sorghum (*Sorghum bicolor* L.) (67), glyphosate-inhibited transpiration in bean (*Phaseolus vulgaris* L.) shoots (68), and glyphosate-inhibited anthocyanin synthesis (69) in buckwheat (*Fagopyrum esculentum* Moench). There are, however, only a few reports [duckweed (61), mouseearcress (*Arabidopsis thaliana* L.) (70), and grain sorghum (67)] in which significant prevention of glyphosate effects on growth has been obtained by feeding intact higher plants supplemental amino acids. Glyphosate inhibition of growth is only marginally prevented, or not prevented at all, by aromatic amino acid feeding in studies with maize (*Zea mays* L.) (39) soybean (71), wheat (*Triticum aestivum* L.) (72) and bean seedlings (73), as well as quackgrass (*Agropyron repens* L. Beauv.) nodes (72). Furthermore, there is little or no effect of glyphosate on aromatic amino acid pools in some cases with higher plant tissue cultures (64) and glyphosate-inhibited growth cannot always be reversed with supplemental amino acids (74).

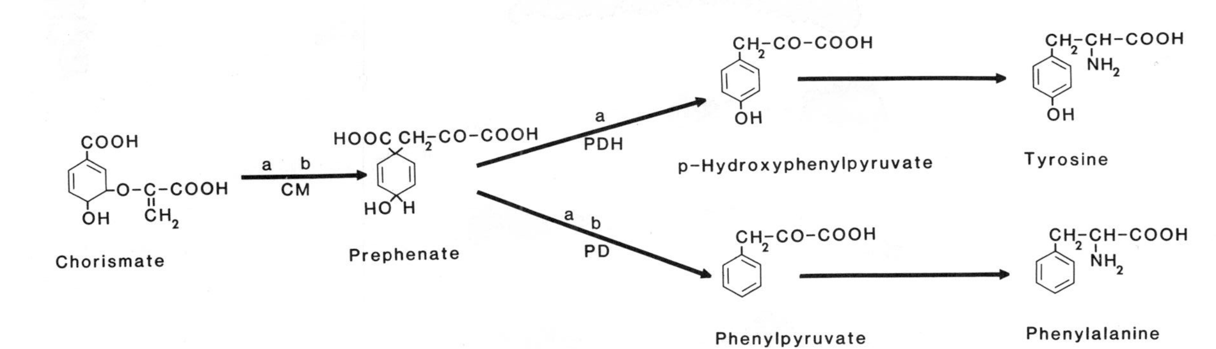

Figure 4. Enzymes of Rhizobium *(a) and* Lemna *(b) proposed as sites of glyphosate inhibition of aromatic amino acid synthesis. Abbreviations: CM, chorismate mutase; PDH, prephenate dehydrogenase; and PD, prephenate dehydratase.*

Table II. Effects of glyphosate on enzymes of phenolic metabolism (see Figure 1 for pathway).

Enzyme	Source	Extractable activity	*In vitro* activity	Reference
3-Deoxy-2-oxo-D-arabino-heptulosate-7-phosphate synthase (EC 4.6.1.3)	*E. coli*	elevated	weak inhibition	(62, 88)
5-Dehydroquinate synthase (EC 4.6.1.3)	*E. coli*	no effect	weak inhibition	(62, 88)
Shikimate dehydrogenase (EC 1.1.1.25)	*Triticum aestivum*	elevated	-	(72)
	Glycine max	no effect	no effect	(71)
Shikimate kinase (EC 2.7.1.71)	*Aerobacter aerogenes*	-	no effect	(87)
5-Enolpyruylshikimate-3-phosphate synthase (EC 2.5.1.19)	*Aerobacter aerogenes*	-	strong inhibition	(86, 87)
	E. coli	-	inhibited	(87)
	Vigna radiata	-	inhibited	(87)
Chorismate synthase (EC 4.6.1.4)	*Aerobacter aerogenes*	-	no effect	(87)
Chorismate mutase (EC 5.4.99.5)	*E. coli*	elevated	none	(62, 88)
	Triticum aestivum	elevated	-	(72)

Table II. (Continued)

Enzyme	Source	Extractable activity	*In vitro* activity	Reference
Prephenate dehydrogenase (EC 1.3.1.12)	*E. coli*	elevated	none	(62, 88)
Prephenate dehydratase (EC 4.2.1.51)	*E. coli*	elevated	none	(62, 88)
Anthranilate synthase (EC 4.1.3.27)	*Aerobacter aerogenes*	-	no effect	(87)
	E. coli	elevated	weak inhibition	(88)
Phenylalanine ammonia-lyase (EC 4.3.1.5)	*Zea mays*	elevated	none	(39, 78)
	Glycine max	elevated	none	(41, 42, 79, 80)
	Gossypium hirsutum	elevated	-	(Duke & Hoagland, unpub.)
	Agropyron repens	elevated	-	(72)
	Triticum aestivum	elevated	-	(72)
	Fagopyrum esculentum	no effect	-	(69)
	Vigna radiata	elevated	-	(Hoagland, unpub.)
Polyphenol oxidase (EC 1.10.3.2)	*Triticum aestivum*	elevated	-	(72)
	Glycine max	elevated	-	(Hoagland & Duke, unpub.)

The effects of glyphosate on several E. coli enzymes of the aromatic amino acid biosynthetic pathway have been studied (62) (Table II). In vitro tests indicated that 3-deoxy-2-oxo-D-arabinoheptonic acid-7-phosphate synthetase (Figure 3, No. 1) and 5-dehydroquinic acid synthetase (Figure 3, No. 2) were inhibited by glyphosate, but only at 10 mM. Both of these inhibitory effects were removed by the addition of Co^{+2}. Chorismate mutase, prephenate dehydrogenase, and prephenate dehydratase (Figure 3, Nos. 7, 8, and 10) were not affected. Herbicide concentrations required for in vitro enzyme effects were higher than apparent physiological levels (i.e., growth-inhibiting concentrations) of glyphosate. These reports indicate that the mode of action hypothesis of Jaworski does not adequately explain the phytotoxic action of glyphosate in all plant systems.

Effects of Glyphosate on PAL. Because of inadequate substantiation of the above theory, we initially postulated that lowered phenylalanine and tyrosine pools caused by glyphosate might additionally be attributed to induction of phenylalanine ammonia-lyase [(PAL) Figure 3, Nos. 14 and 15] activity. There was some evidence from other reports that high PAL activity could retard growth through aromatic-amino acid depletion (75). PAL deaminates tyrosine to some extent in most plant systems (76), thus, increased PAL activity could result in less-than-adequate levels of phenylalanine and tyrosine required for normal protein synthesis. PAL, by acting on phenylalanine, plays a key role in phenylpropanoid biosynthesis and regulates the formation of a variety of phenolic compounds (Figure 3). This enzyme has been shown to be regulated by a number of environmental factors (water stress, wounding, infection, light, chemical action, etc.) (76). Accumulation of phenolics could cause an autoallelopathic or phytotoxic effect on the plants. Furthermore, the non-oxidative deamination of phenylalanine could yield toxic levels of ammonia if deamination enzymes did not provide protection (Figure 5). In tests of this hypothesis with maize and soybean seedlings, we found that glyphosate caused pronounced increases in extractable PAL activity (39, 41, 42) (Figures 6 and 11). In vitro tests showed that glyphosate had no direct effect on PAL. Later, Cole et al. (72) found that glyphosate increased PAL activity in single node buds of quackgrass rhizomes and in root tips of wheat (Table III). Holländer and Amrhein (69), however, found no effect of glyphosate on extracted PAL of buckwheat hypocotyls. We found a good correlation between glyphosate-caused PAL activity increases, and substrate (phenylalanine) and product (hydroxyphenolic) decreases caused by glyphosate in maize (Figure 7) and soybean seedlings (Figure 8). Nilsson's laboratory also showed that glyphosate decreased aromatic amino acid levels in wheat roots (77). Phenolic levels were

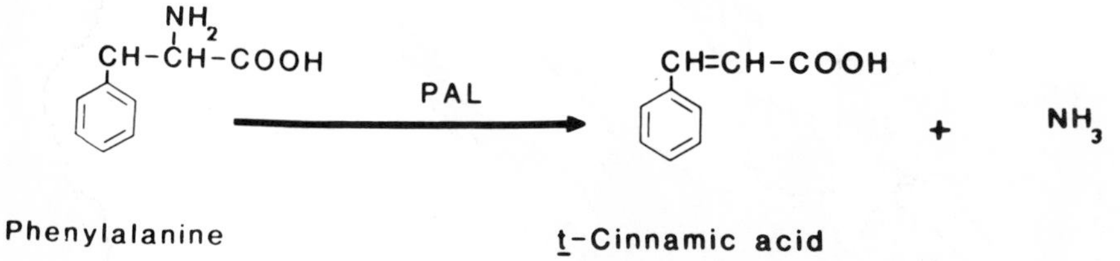

NH$_2$

CH–CH–COOH → (PAL) → CH=CH–COOH + NH$_3$

Phenylalanine t-Cinnamic acid

Figure 5. Nonoxidative deamination of phenylalanine by PAL.

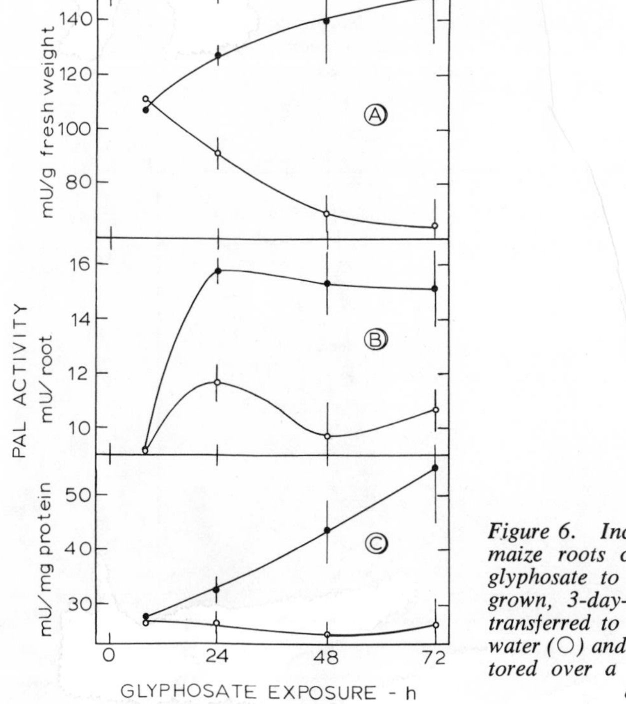

Plant Science Letters

Figure 6. Increased extractable PAL in maize roots caused by root-feeding of glyphosate to intact plants (39). Dark-grown, 3-day-old maize seedlings were transferred to 1 mM glyphosate (●), or water (○) and enzyme activity was monitored over a 3-day time course during dark growth.

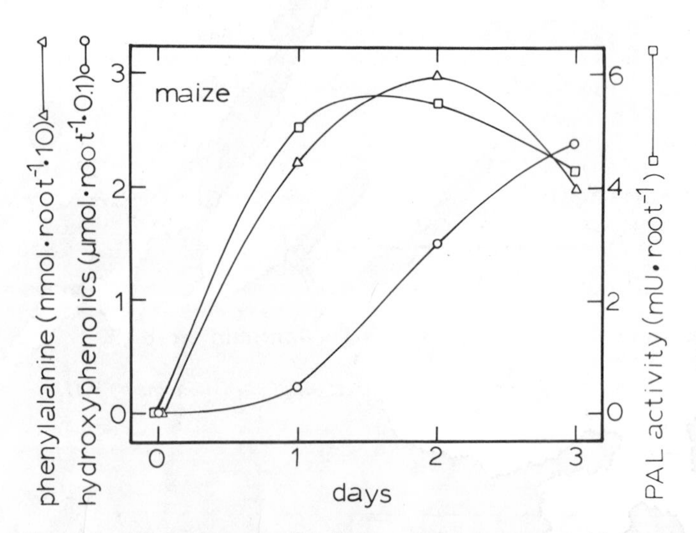

Figure 7. Correlation of extractable PAL activity increases with decreases in PAL substrate (phenylalanine) and products (hydroxyphenolics) during glyphosate treatment in roots of maize seedlings (39, 78).

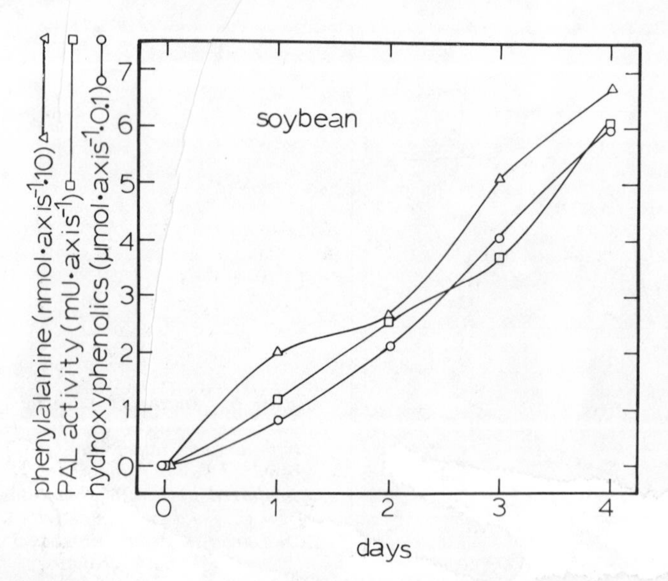

Figure 8. Correlation of extractable PAL activity increases with decreases in PAL substrate (phenylalanine) and products (hydroxyphenolics) during glyphosate treatment in axes of soybean seedlings (41, 42, 80).

generally higher on a fresh weight basis but were lower on a plant organ basis in glyphosate-treated tissues (39, 41, 42, 78). Although several equivocations can be made concerning the hydroxyphenolic data due to techniques used (78), the lowering of anthocyanin levels (79) indicates that these results are qualitatively accurate. The results of our initial studies (39, 41, 42, 78) indicated that although glyphosate has profound effects on extractable PAL, PAL substrate(s), and PAL end products, increased PAL activity was probably a secondary effect of decreased feedback control due to decreased substrate and, thus, decreased product.

Table III. Effect of 0.5 mM glyphosate on extractable activity of four enzymes of phenolic metabolism from wheat root tips. Adapted from Cole *et al.* (72).

	Enzyme activity (units mg^{-1} protein) after 24 h	
Enzyme	Control	Glyphosate
Chorismate mutase	2.3	5.9
Shikimate dehydrogenase	4.5	11.5
PAL	3.3	19.1
Polyphenol oxidase	6.1	14.1

We conducted further experiments to determine if glyphosate's effects could be reversed in higher plants by increasing aromatic amino acid levels using PAL inhibitors (80) (Figure 9). We reasoned that if increased PAL activity was involved in glyphosate's mode of action, then blocking PAL *in vivo*, with a PAL inhibitor might reduce or reverse glyphosate's toxic effects. The PAL inhibitor, α-aminooxy-β-phenylpropionic acid (AOPP) had previously been shown to effectively inhibit PAL and to inhibit anthocyanin accumulation while having little or no effect on growth (81, 82). At 0.1 mM, AOPP had no significant effect on soybean growth until 96 hours, but did provide marginal growth reversal (10%) of 0.5 mM glyphosate's inhibition (80) (Figure 10). This level of AOPP increased phenylalanine and tyrosine levels in glyphosate-treated tissues to control levels. PAL activity from axes was completely inhibited *in vitro* by 10 μM AOPP. Extractable PAL activity was increased by AOPP (Figure 11) by all measurement criteria. This increase was probably because of decreased feedback inhibition of PAL by its products and

α-Aminooxyacetate ; AOA NH_2-O-CH_2-COOH

α-Aminooxy-β-phenylpropionate ; AOPP $NH_2-O-CH(CH_2C_6H_5)-COOH$

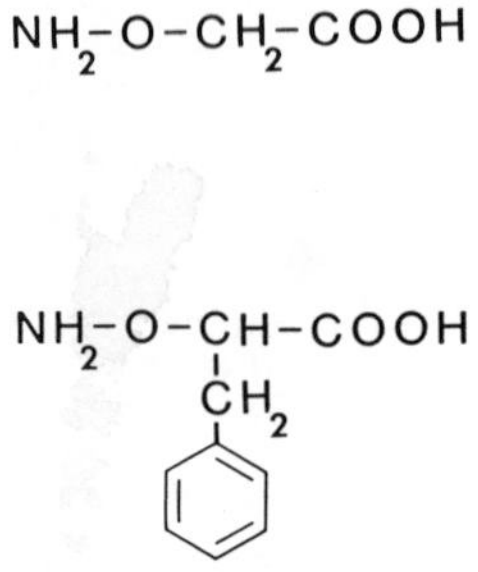

Figure 9. PAL inhibitors, α-aminooxy-β-phenylpropionic acid (AOPP) and aminooxyacetic acid (AOA).

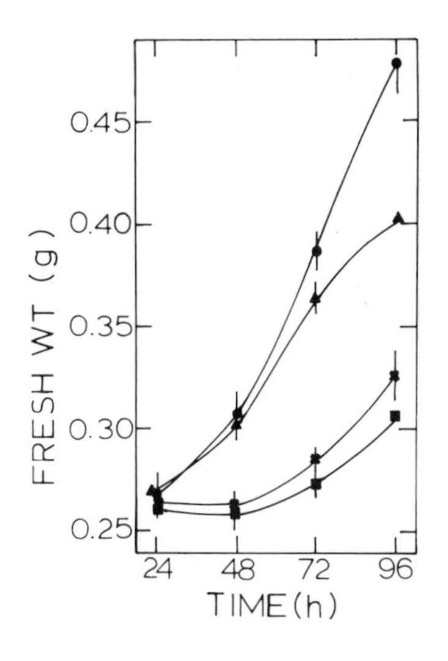

Plant Physiology

Figure 10. The effect of AOPP and glyphosate on soybean axis growth when supplied to roots in liquid culture (80). Seedlings were exposed to continuous white light and root-fed various chemicals after 3 days of dark growth. Key: ●, *control;* ■, *0.5 mM glyphosate;* ▲, *0.1 mM AOPP; and* ✖, *glyphosate (0.5 mM) plus AOPP (0.1 mM).*

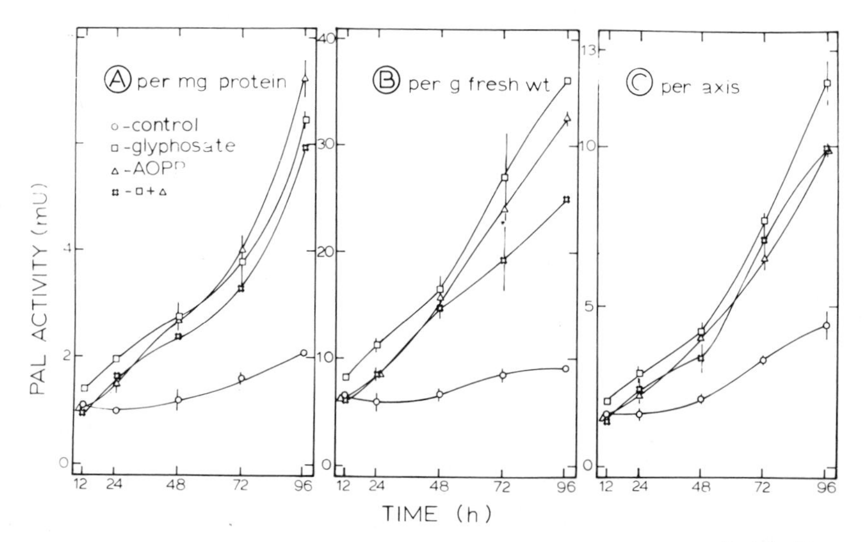

Plant Physiology

Figure 11. The effect of AOPP and glyphosate on extractable PAL activity in light-grown soybean axes (80). Seedlings were exposed to continuous white light and root-fed various chemicals after 3 days of dark growth. Key: ○, *control;* □, *0.5 mM glyphosate;* △, *0.1 mM AOPP; and* ⌑, *glyphosate (0.5 mM) plus AOPP (0.1 mM).*

the fact that our purification and extraction procedures for PAL removed AOPP. AOPP, glyphosate, and AOPP plus glyphosate reduced total hydroxyphenolic accumulation. Similar experiments were performed with aminooxyacetic acid (Figure 9). A summary of the effects of these two PAL inhibitors and glyphosate on growth and metabolism of soybean seedlings is presented in Table IV. Cole _et al._ (_72_) found no reversal of glyphosate effects with the PAL inhibitors, D-phenylalanine, cinnamic acid, and AOPP. From these data, it can be concluded that glyphosate's enhancement of PAL activity plays a role in depletion of phenylalanine and tyrosine, but that PAL's role in glyphosate's mode of action, taken alone, may not be of major importance in its phytotoxic effects. Moreover, these results strongly suggest that interference with aromatic amino acid metabolism does not fully explain glyphosate's mode of action in higher plants.

Table IV. Summary of effects of glyphosate (0.5 mM) and two PAL inhibitors, AOPP (0.01 mM) and AOA (0.01 mM), on growth and metabolism of soybean seedlings.

Treatment	Growth	Soluble Protein	Soluble Phenolics	Extracted PAL	Phenyl-alanine
Glyphosate	-	-	-	+	-
AOPP	0	0	-	+	+
AOA	-	-	-	±	+

These conclusions are supported by studies in our laboratory in which we found only a minimal reversal of glyphosate-caused growth inhibition by root-fed aromatic amino acids (_71_). Feeding aromatic amino acids before glyphosate exposure did not enhance reversal. On a fresh-weight basis, glyphosate had no inhibitory effect on uptake or incorporation of these amino acids into protein or secondary phenolics. These data suggest that either root-fed aromatic amino acids are compartmentalized differently than the endogenous pools that are affected by glyphosate, or that root-fed glyphosate exerts most of its effect on soybean growth through means other than inhibition of aromatic amino acid biosynthesis.

Effects of Glyphosate Analogs on Secondary Phenolic Compound Synthesis. To test the specificity of glyphosate's action and/or the effect of analogs and possible degradation products (Figure 2), a study was conducted that used these compounds on the soybean system (79). Glyphosine was not as effective in increasing activity of PAL as was glyphosate. The other compounds had no effect, except for aminomethylphosphonic acid, which slightly reduced extractable PAL activity. Hydroxyphenolic levels were analyzed, but only glyphosine caused a decrease in phenolics and the other compounds had no effect. Anthocyanin levels were reduced by 50% in glyphosate-treated seedlings, but were decreased to a much lesser degree by glyphosine. Holländer and Amrhein (69) obtained almost identical results with buckwheat hypocotyls. Glyphosine was generally not as effective in inhibiting chlorophyll accumulation as was glyphosate. Thus, glyphosate's effect on PAL and the other parameters are rather specific when compared to those of its structural analogs.

Specificity of Glyphosate's Effect on PAL. Glyphosate's effects on PAL are rather specific when compared to other herbicides. The effects of 16 herbicides representing 14 herbicide classes on extractable PAL activity from both light- and dark-grown soybean seedling tissues were determined (83). Only 2 compounds increased extractable PAL activity: amitrole (3-amino-*s*-triazole), to about 30% of that of glyphosate on a per axis basis; and paraquat (1,1'-dimethyl-4,4'-bipyridinium ion), which showed a somewhat greater enhancement at early exposure times while later lowering extractable PAL levels. These results support the view that glyphosate-caused PAL increases are specific and are not a secondary effect of stress.

Effects of Glyphosate on Enzymes of Aromatic Amino Acid Synthesis. Although early work on glyphosate effects on enzymes of aromatic amino acid synthesis were inconclusive (62), recent work in Amrhein's laboratory has revealed the apparent site of action of this compound. They found that anthocyanin synthesis in illuminated, excised hypocotyls of buckwheat was severely depressed by glyphosate (69). This occurred when glyphosate was applied *via* root uptake, to floated excised hypocotyls, or by spraying seedlings. L-phenylalanine was the only aromatic amino acid that effectively reversed inhibition of anthocyanin synthesis. When anthocyanin synthesis was allowed to proceed for 10 h and then continued in the presence of 3 mM glyphosate, its rate was reduced within less than 1 h. They suggested that this rapid inhibition was due to a direct inhibition by glyphosate of a metabolic step in the pathway leading to anthocyanin rather than to the induced appearance (or disappearance) of an enzyme. Investigation of the rates of inhibition of chlorophyll and anthocyanin formation by glyphosate indicated that inhibition of anthocyanin formation was more sensitive than that of chlorophyll by an order of magnitude (Figure 12).

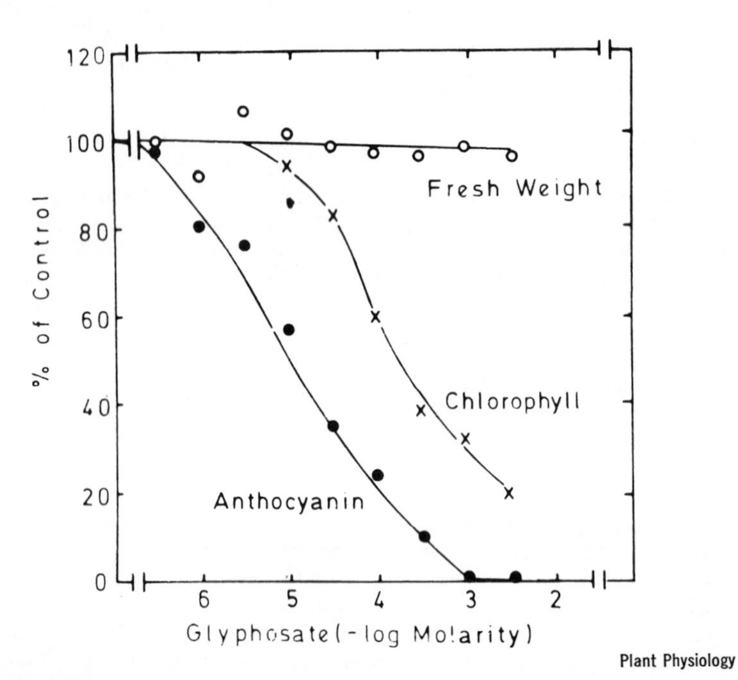

Figure 12. Effect of glyphosate on fresh weight increase, chlorophyll content, and anthocyanin content in excised buckwheat cotyledons (69). Cotyledons of 6-day-old etiolated seedlings were incubated at the indicated glyphosate concentrations for 24 h in the light.

Although phenylalanine alone was able to alleviate glyphosate inhibition of anthocyanin formation, feeding tyrosine in addition to phenylalanine was required to achieve partial alleviation of the inhibition of chlorophyll formation. Additional tryptophan did not further increase the chlorophyll content of cotyledons. Further work with buckwheat indicated that a 24-h light or dark incubation of untreated excised hypocotyls had little effect on endogenous shikimate content (84). Glyphosate, however, (1 mM) caused about a 20-fold increase in the shikimate concentration in darkness and a greater than 50-fold increase in the light. The glyphosate and light treatment raised the shikimate concentration in the tissue to nearly 2 mM. Greater glyphosate-caused shikimate accumulation in the light than in the dark indicated that light increased aromatic amino acid synthesis (85). Glyphosate, even at 1 mM, did not inhibit the growth of *G. mollugo* cells in a modified B5 medium if the medium was fortified with casein hydrolysate (84). In the absence of exogenous amino acids, 0.3 mM glyphosate inhibited cell growth by 55%. Concentrations of glyphosate higher than 0.1 mM inhibited anthraquinone production and produced an enormous accumulation of shikimate in the cells (Figure 13). Glyphosate's inhibition of anthraquinone production was partially alleviated by 1 mM chorismate and 1 mM *o*-succinyl-benzoate, but not by 1 mM phenylalanine or tyrosine, either alone or in combination. Chorismate alone, and the combination of phenylalanine and tyrosine, inhibited anthraquinone formation slightly, whereas *o*-succinyl-benzoate increased pigment formation. Glyphosate was found to be a powerful inhibitor of the formation of anthranilate from shikimate; 50% inhibition was achieved with 5 to 7 mM exogenous glyphosate. Glyphosine, aminomethylphosphonate and iminodiacetate showed no inhibitory activity.

Additional work in Amrhein's laboratory (86, 87) investigated the effect of glyphosate on the conversion of shikimate to anthranilate. The activities of shikimate kinase (Figure 3, No. 4), 5-enolpyruvylshikimate-3-phosphate synthase (Figure 3, No. 5), chorismate synthase (Figure 3, No. 6), and anthranilate synthase (Figure 3, No. 12) in cell-free extracts of *Aerobacter* were studied. Of the four enzymes involved in this transformation, only 5-enolpyruvylshikimate-3-phosphate synthase was inhibited by glyphosate (Table II, Figure 14). A highly significant correlation between the accumulation of shikimate and reduction of anthocyanin formation in buckwheat hypocotyls in the presence of various concentrations of glyphosate was found. Shikimate-3-phosphate was identified as the product that accumulated in a cell-free glyphosate-treated system that enzymatically converted shikimate to anthranilate.

Roisch and Lingens (88) recently found 3-dehydroquinate synthase (Figure 3, No. 2-alternate nomenclature) and phospho-2-oxo-3-deoxyhepton-acetaldolase (Figure 3, No. 1-alternate

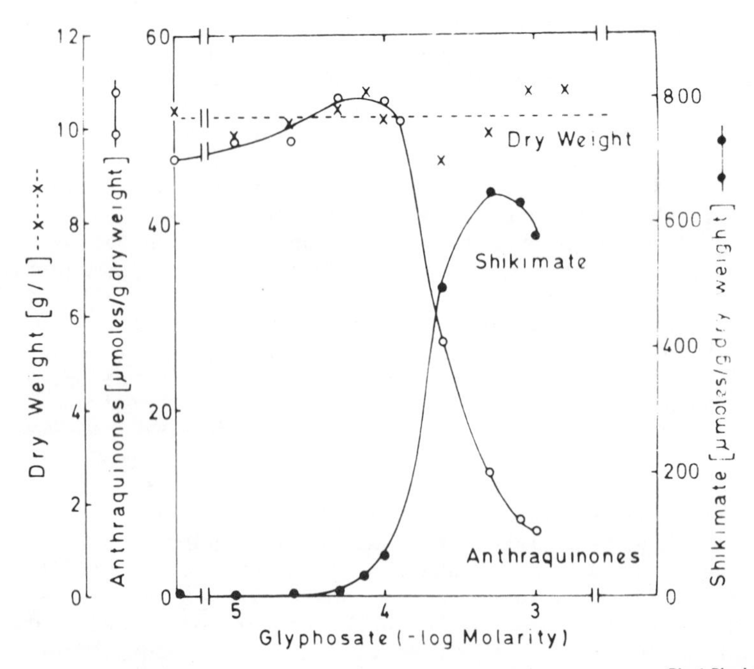

Plant Physiology

Figure 13. Glyphosate effect on anthraquinone production and shikimate accumulation in G. mollugo *cell cultures (84). Cells were harvested 10 days after inoculation with 2 cm³ of packed cells into 25 mL medium containing glyphosate. Anthraquinone content of the cells was corrected for the amount present in the inoculum.*

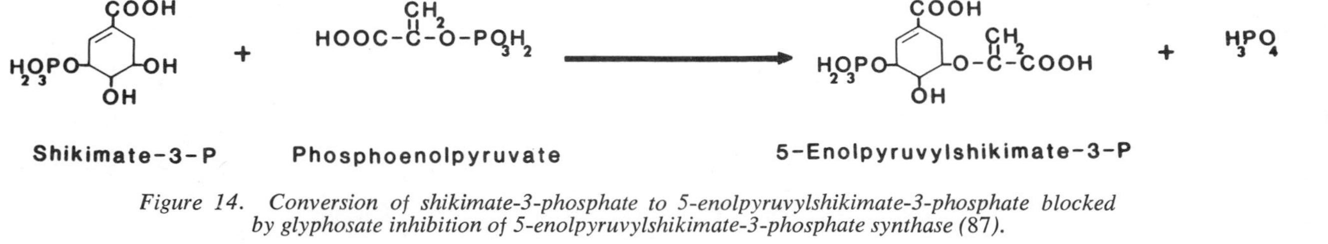

Figure 14. Conversion of shikimate-3-phosphate to 5-enolpyruvylshikimate-3-phosphate blocked by glyphosate inhibition of 5-enolpyruvylshikimate-3-phosphate synthase (87).

nomenclature) to be inhibited by glyphosate at high (e.g., 10 mM), non-physiological concentrations. They found extracted activities of several enzymes of aromatic amino acid synthesis from *E. coli* to increase in response to glyphosate treatment.

Cole *et al.* (72) found that glyphosate increased extracted specific activities of two shikimate pathway enzymes, shikimate dehydrogenase (shikimate: NADP oxidoreductase, Figure 3, No. 3) and chorismate mutase (Figure 3, No. 7), as well as PAL activity (Table III). Polyphenol oxidase (Figure 3, No. 16) activity was also increased, although not to as great an extent as PAL activity. In our work (71), we found no effects on shikimate dehydrogenase.

The ability of glyphosate to block phenylalanine synthesis, resulting in shikimate accumulation, makes this herbicide a useful tool for examining control of secondary metabolism. Berlin and Witte (89) have determined that PAL, rather than substrate supply, is most important in regulating cinnamoyl putrescine synthesis in tobacco cell cultures.

Effects of Glyphosate on the Chloroplast. Glyphosate is generally not thought to have a direct effect on photosynthesis, although Van Rensen (90) reported that glyphosate inhibits PSII electron transport in spinach (*Spinacia oleracea* L.) chloroplasts as well as O_2 evolution in *Scenedesmus*. Richard *et al.* (91) showed that what Van Rensen observed was probably a pH effect. Using the closely related compound, glyphosine, Croft *et al.* (92) found no effect on photosynthetic electron transport or cyclic photophosphorylation. Sprankle *et al.* (17) concluded that glyphosate's effects on photosynthesis are secondary because effects on photosynthetic rates were not measureable until 72 h after treatment. Shaner and Lyon (93), however, measured effects on photosynthesis 6 h after treatment.

Although there appear to be no direct effects on photosynthesis, glyphosate may directly inhibit chlorophyll synthesis. Glyphosate was shown to strongly inhibit chlorophyll accumulation (94) as well as incorporation of labeled precursors into α-aminolevulinic acid (95) (Table V). Glyphosate did not, however, inhibit incorporation of labeled α-aminolevulinic acid into chlorophyll. Yet, in systems in which glyphosate effects on chlorophyll and anthocyanin accumulation were simultaneously compared, anthocyanin synthesis was more (69) or equally sensitive to glyphosate (79). Abu-Irmaileh and Jordan (26) found that although chlorosis was an early symptom of glyphosate injury in purple nutsedge (*Cyperus rotundus* L.), carotenoids were more strongly affected. Levels of catalase, another porphyrin-containing compound, were most rapidly reduced by glyphosate, however. This suggests that the effect on chlorophyll may be indirect through photobleaching and/or peroxidation of chlorophyll. In maize seedlings, however, Ali and Fletcher (96) found that chlorophyll content was more strongly affected by glyphosate than was carotenoid content.

Table V. Incorporation of ^{14}C-labeled precursors of 5-aminolevulinic acid (ALA) into ALA of 7-d-old etiolated barley shoots treated with 1 mM glyphosate, incubated in total darkness for 8 h, and then illuminated for 13 h. From Kitchen *et al.* (95). (Reproduced by permission of the Weed Science Society of America)

Treatment	^{14}C-ALA	Reduction
	(dpm/g)	(%)
^{14}C-Glutamate		
Control	2174	76.9
Glyphosate	503	
^{14}C-α-Ketoglutarate		
Control	530	92.1
Glyphosate	42	
^{14}C-Glycine		
Control	480	91.3
Glyphosate	42	

Ultrastructural Analysis of Glyphosate Effects. Few studies of ultrastructural effects of glyphosate have been published. Campbell *et al.* (28) reported ultrastructural damage to plastids of quackgrass mesophyll cells as early as 24 h after glyphosate treatment. They found disruption of the chloroplast envelope, plastoglobuli increases, swelling of the rough endoplasmic reticulum, and cytoplasmic vesicle formation. The effects on plastids were no more severe than effects on other organelles. Very similar effects of glyphosate have been noted on the ultrastructure of white mustard (*Sinapis alba* L.) seedlings (97). In the liverwort *Pellia epiphylla*, Pihakaski and Pihakaski (98) found that glyphosate caused vesicular and tubular structures on the chloroplast surface as well as in the cytoplasm. They also described granular bodies, increases in lipid spherules and cytoplasm vacuolation, and deterioration of oil bodies, endoplasmic reticula, and ribosomes in the glyphosate-treated plant. Generally, ultrastructural studies of glyphosate effects have not been supplemented with biochemical and physiological investigations.

Membrane Transport. Glyphosate apparently has little effect on efflux of cellular contents except at very high concentrations (99-102). Uptake of ^{86}Rb and ^{32}P was found to be retarded in isolated bean (*Phaseolus vulgaris* L.) cells before effects on photosynthesis, respiration, and RNA and protein synthesis (99). This effect was not due to loss of membrane integrity, decrease in energy supply, or external ion chelation. These results strongly suggest that more research should be conducted on glyphosate effects on metal ion uptake and utilization.

Glyphosate has been found to retard uptake of amino acids, nucleotides, and glucose in isolated cells (99). Other studies have indicated that glyphosate-caused inhibition of uptake of amino acids by intact plant roots (71) or excised buds (72) is not severe.

Respiration Effects. Glyphosate has been reported to uncouple oxidative phosphorylation in plant (103) and mammalian (104) mitochondria. Respiratory activity in maize roots was found to be reduced 6 h after foliar application of glyphosate (96). Tetrazolium reduction was greatly reduced in wheat root tips by 0.5 mM, but not by 0.1 mM glyphosate after 1 day (72). In wheat and in quackgrass, however, glyphosate was shown to affect photosynthesis much earlier and more severely than respiration in leaf tissues (17). Brecke and Duke (99) found no effect on O_2 consumption by isolated bean cells.

Chelation Studies. Amino-phosphonic acids such as glyphosate chelate metal cations in aqueous media (11). This property could affect plant metabolism and physiology at any of the many points dependent on metal cations. The effects of glyphosate on aromatic amino acid synthesis in *E. coli* has been attributed to chelation of Co^{+2} and Mg^{+2} (88), cofactors for enzymes in this pathway. Addition of these cations to the media stopped the inhibition of two enzymes of aromatic amino acid synthesis. Holländer and Amrhein (69) found no effect of Al^{+3}, Fe^{+2}, Co^{+2}, or Ca^{+2} on glyphosate-decreased anthocyanin synthesis. Hensley *et al.* (13) found that $FeCl_3$, $FeCl_2$, and $AlCl_3$ reduced the activity of glyphosate, but that $CaCl_2$, KCl and NaCl had no effect. Sprankle *et al.* (17), however, found $FeSO_4$ to have no effect on glyphosate toxicity. Gresshoff (63) found that Zn^{+2}, Co^{+2}, and Fe^{+2} had no effect on glyphosate toxicity to unicellular organisms.

Other Metabolic Effects. Any compound which greatly retards growth or causes death will ultimately affect all cellular processes. Many apparent secondary effects of glyphosate have been examined. Retardation of protein synthesis is apparently one such effect. Cole *et al.* reported that incorporation of ^{14}C-labeled leucine was inhibited more by glyphosate than incorporation of labeled phenylalanine, suggesting that protein synthesis is slowed by depletion of aromatic amino acid synthesis (72). In their system, glyphosate caused decreased soluble protein levels. In isolated bean cells, however, Brecke and Duke (99) found inhibition of incorporation of labeled leucine into protein to be explained fully by inhibition of uptake of the label. On a fresh weight basis, Duke and Hoagland (71) found no effect of growth-retarding levels of glyphosate on uptake and incorporation of aromatic amino acids into proteins of soybean seedlings, although relatively more of these amino acids were incorporated into protein in the control than in the glyphosate treatment. That glyphosate induces prolonged increases in extractable levels of PAL, increases in activities of other enzymes of phenolic biosynthesis, causes transient increases in nitrate reductase in some tissues (Hoagland unpublished), and has no effects on levels of the protein-pigment phytochrome (41) suggest that glyphosate's effects on protein synthesis are not primary.

Glyphosate may affect the synthesis of non-aromatic amino acids. Nilsson (77) suggested that the build-up of glutamate and glutamine in glyphosate-treated tissue might be due to blocked transamination reactions. No studies have been reported of the effect of glyphosate on transamination reactions, but numerous changes in free amino acid profiles are caused by glyphosate (41, 42, 66, 77, 78). Methionine levels are greatly reduced by glyphosate (42), which suggests that this herbicide may alter ethylene synthesis. Results of Baur (105) suggest that glyphosate may inhibit auxin transport by increasing ethylene synthesis. Ethylene is also known to increase extractable PAL levels (76). Lee (74) showed that glyphosate effects on growth of tobacco and soybean callus cultures could be prevented by addition of indoleacetic acid.

Summary and Conclusions

Separation of primary herbicidal effects from the secondary, tertiary, or quaternary effects on plants is difficult. Metabolic processes are interdependent *in vivo* and thus it is difficult to extrapolate with certainty *in vitro* data to a living system. Thus, while the primary site of action appears to be known for some herbicides, for many others the primary action mechanism remains obscure. The mechanism of action of some herbicides may not be determined until further elucidation of the biochemical parameters of both major and minor plant processes.

Considerable biochemical data on glyphosate action and its effects in plants have been accumulated, but more data are needed. Glyphosate's rapid absorption and translocation to various plant organs and the lack of metabolism in plant tissues suggest that it may come in contact with and alter a wide variety of functions and enzymes and thus may have multiple sites of action. Whether glyphosate's mechanism of action in higher plants can be explained entirely by its effects on aromatic amino acid synthesis and phenolic compound metabolism is not yet known. We suspect that although this effect is partially involved in glyphosate's mechanism of action in intact higher plants, another crucial process(es) (not directly related to phenolic metabolism or necessary for tissue culture growth) is directly affected by glyphosate. Glyphosate's properties as a divalent metal cation chelator suggest avenues of research that have not been sufficiently explored. Additionally, more biochemical and physiological information on analog effects, structure-function relationships, and antidotes is needed for a clearer understanding of glyphosate's mechanism of action and to provide the groundwork for altering specificity and selectivity of this unique herbicide class.

Literature Cited

1. Baird, D.D.; Upchurch, R.P.; Homesley, W.B.; Franz, J.E. Proc. North Centr. Weed Contr. Conf. 1971, 26, 64-8.
2. Serdy, F.S. Weeds Today 1975, 6, 19-20.
3. Fischer, B.B.; Swanson, F.H.; May, D.M.; Tange, A.H. Agrichem. Age 1976, 19, 16-21.
4. Carlsson, T. Swedish Weed Contr. Conf. Proc. 1978, 1, 1-4.
5. Franz, J.E. "Advances in Pesticide Science"; Geissbüehler, H., Ed.; Part 1; Pergamon Press: Oxford, 1979; pp 139-47.
6. Chykaliuk, P.B.; Abernathy, J.R.; Gipson, J.R. Texas Agric. Exp. Stat. Public. 1979, MP 1443, 87 pp.
7. Herbicide Handbook of the Weed Sci. Soc. of America, 4th Ed., Champaign, IL, 1979; p. 479.
8. Hoagland, R.E.; Duke, S.O. Weeds Today 1981, 12, 21-3.
9. Hoagland, R.E.; Duke, S.O. Residue Rev., in preparation.
10. Ashton, F.M.; Crafts, A.S. "Mode of Action of Herbicides". 2nd Ed., John Wiley and Sons: New York, 1981 236-53.
11. Kabachnik, M.I.; Medved, T.Y.; Dyatova, N.M.; Rudomino, M.V. Russ. Chem. Rev. 1974, 43, 733-44.
12. Turner, D.J.; Loader, M.P.C. Weed Res., 1978, 18, 199-207.
13. Hensley, D.L.; Beverman, D.S.N.; Carpenter, D.L. Weed Res. 1978, 18, 289-91.
14. Carter, R.P.; Carrol, R.L.; Ivan, R.R. Inorg. Chem. 1967, 6, 939-42.

15. Segura, J.; Bingham, S.W.; Foy, C.L. Weed Sci. 1978, 26, 32-6.
16. Putnam, A.R. Weed Sci. 1976, 24, 425-30.
17. Sprankle, P.; Meggitt, W.F.; Penner, D. Weed Sci. 1975, 23, 235-40.
18. Claus, J.S.; Behrens, R. Weed Sci. 1976, 24, 149-52.
19. Zandstra, C.H.; Nishimoto, R.K. Weed Sci. 1977, 25, 268-74.
20. Chase, R.L.; Appleby, A.P. Weed Res. 1979, 19, 241-6.
21. Gothrup, O.; O'Sullivan, P.A.; Schraa, R.J.; Vanden Born, W.H. Weed Res. 1976, 16, 197-201.
22. Moshier, L.; Penner, D. Weed Sci. 1978, 26, 686-91.
23. Hildebrand, L.D.; Sullivan, D.S.; Sullivan, T.P. Bull. Environ. Contam. Toxicol. 1980, 25, 353-7.
24. Sullivan, D.S.; Sullivan, T.P.; Bisalputra, L. Bull. Environ. Contam. Toxicol. 1981, 26, 91-6.
25. Suwannamek, U.; Parker, C. Weed Res. 1975, 15, 13-9.
26. Abu-Irmaileh, B.E.; Jordan, L.S. Weed Sci. 1978, 26, 700-3.
27. Fernandez, C.H.; Bayer, D.E. Weed Sci. 1977, 25, 396-400.
28. Campbell, W.F.; Evans, J.O.; Reed, S.C. Weed Sci. 1976, 24, 22-5.
29. Lutman, P.J.W.; Richardson, W.G. Weed Res. 1978, 18, 65-70.
30. Marriage, P.B.; Khan, S.U. Weed Sci. 1978, 26, 374-8.
31. Clay, D.V. Proc. Brit. Weed Contr. Conf. 1972, 11, 451-7.
32. Curtis, O.F. Proc. Northeast. Weed Sci. Soc. 1974, 28, 219.
33. Lange, A.H.; Fischer, B.B.; Elmore, C.L.; Kempen, H.M.; Schesselman, J. Calif. Agric. 1975, 19, 6-7.
34. Abu-Irmaileh, B.E.; Jordan, L.S. Weed Sci. 1977, 30, 57-63.
35. Moosavi-Nia, H.; Dore, J. Weed Res. 1979, 19, 137-43.
36. Moosavi-Nia, H.; Dore, J. Weed Res., 1979, 19, 321-7.
37. Blair, A.M. Weed Res. 1975, 15, 83-8.
38. Wyrill, J.P.; Burnside, O.C. Weed Sci. 1977, 25, 275-87.
39. Duke, S.O.; Hoagland, R.E. Plant Sci. Lett. 1978, 11, 185-90.
40. Haderlie, L.C.; Slife, F.W.; Butler, H.S. Weed Res. 1978, 18, 269-73.
41. Hoagland, R.E.; Duke, S.O.; Elmore, C.D. Physiol. Plant. 1979, 46, 357-66.
42. Duke, S.O.; Hoagland, R.E.; Elmore, C.D. Physiol. Plant. 1979, 46, 307-17.
43. Caseley, J. Proc. Brit. Weed Contr. Conf. 1972, 2, 641-7.
44. Upchurch, R.P.; Baird, D.D. Proc. West. Weed Sci. Soc. 1972, 25, 41-5.

45. Evans, D.M. Proc. Brit. Weed Contr. Conf. 1972, 1, 64-70.
46. Davis, N.E. Ph.D. Thesis, Univ. Wisconsin, Madison. 1976.
47. Schultz, M.E.; Burnside, O.C. Weed Sci. 1980, 28, 13-20.
48. Moshier, L.; Turgeon, A.J.; Penner, D. Weed Sci. 1976, 24, 445-8.
49. Moshier, L.; Penner, D. Weed Sci. 1978, 26, 163-6.
50. Egley, G.H.; Williams, R.D. Weed Sci. 1978, 26, 249-51.
51. Klingman, D.L.; Murray, J.J. Weed Sci. 1976, 24, 191-3.
52. Hance, R.J. Pestic. Sci. 1976, 7, 363-6.
53. Wyrill, J.B.; Burnside, O.C. Weed Sci. 1976, 24, 557-66.
54. McWhorter, C.G. Weed Sci. 1980, 28, 113-8.
55. Jordan, T.N. Weed Sci. 1977, 25, 448-51.
56. Davis, H.E.; Fawcett, R.S.; Harvey, R.G. Weed Sci. 1979, 27, 110-4.
57. Wills, G.D. Weed Sci. 1978, 26, 509-12.
58. McWhorter, C.G.; Jordan, T.N.; Wills, G.D. Weed Sci. 1980, 28, 113-8.
59. Ahmadi, M.S.; Haderlie, L.C.; Wicks, G.A. Weed Sci. 1980, 28, 277-82.
60. Ruppel, M.; Brightwell, B.B.; Schaefer, J.; Marvel, J.T. J. Agric. Food Chem. 1977, 25, 517-28.
61. Jaworski, E.G. J. Agric. Food Chem. 1972, 20, 1195-8.
62. Roisch, V.; Lingens, F. Angew. Chem. 1974, 13, 400.
63. Gresshoff, P.M. Aust. J. Plant Physiol. 1979, 6, 177-85.
64. Haderlie, L.C.; Widholm, J.M.; Slife, F.M. Plant Physiol. 1977, 60, 40-3.
65. Haderlie, L.C. Ph. Thesis, Univ. of Ill. 1975, 172 pp.
66. Tymonko, J.M; Foy, C.L. Plant Physiol. Suppl. 1978, 61, 41.
67. Baur, J.R. Weed Sci. 1979, 27, 69-72.
68. Shaner, D.L.; Lyon, J.L. Weed Sci. 1980, 28, 31-5.
69. Holländer, H.; Amrhein, N. Plant Physiol. 1980, 66, 823-9.
70. Gresshoff, P.M. Arabidopsis Inf. Serv. 1979, 16, 73.
71. Duke, S.O.; Hoagland, R.E. Weed Sci. 1981, 29, 297-302.
72. Cole, D.J.; Dodge, A.D.; Caseley, J.C. J. Exp. Bot. 1980, 31, 1665-74.
73. Brecke, B.J. Ph.D. Thesis, Cornell Univ. 1976, 127 pp.
74. Lee, T.T. Weed Res. 1980, 20, 365-9.
75. James, D.J.; Davidson, A.W. Ann. Bot. 1976, 40, 957-68.
76. Camm, E.L.; Towers, G.H.N. "Progress in Phytochemistry", Vol. 4; Reinholt, L.; Harborne, J. B.; Swain, T., Eds.; Pergamon Press: New York, 1977; pp. 169-88.
77. Nilsson, G. Swedish J. Agric. Res. 1977, 7, 153-7.
78. Hoagland, R.E.; Duke, S.O.; Elmore, C.D. Plant Sci. Lett. 1978, 13, 291-9.
79. Hoagland, R.E. Weed Sci. 1980, 28, 393-400.
80. Duke, S.O.; Hoagland, R.E.; Elmore, C.D. Plant Physiol. 1980, 65, 17-21.

81. Amrhein, N.; Gödeke, K.-H. Plant Sci. Lett. 1977, 8, 313-7.
82. Amrhein, N.; Holländer, H. Planta 1979, 144, 385-9.
83. Hoagland, R.E.; Duke, S.O. Weed Sci. 1981, 29, 433-9.
84. Amrhein, N.; Deus, B.; Gehrke, P.; Steinrücken, H.C. Plant Physiol. 1980, 66, 830-4.
85. Amrhein, N.; Holländer, H. Naturwiss. 1981, 68, 43.
86. Amrhein, N.; Shab, J.; Steinrücken, H.C. Naturwiss. 1980, 67, 356-7.
87. Steinrücken, H.C.; Amrhein, N. Biochem. Biophys. Res. Comm. 1980, 94, 1207-12.
88. Roisch, V.; Lingens, F. Hoppe-Seyler's Z. Physiol. Chem. 1980, 361, 1049-58.
89. Berlin, J.; Witte, L. Z. Naturforsch. 1981, 36c, 210-4.
90. Van Rensen, J.J.S. "Proceeding of the Third International Congress on Photosynthesis", Arvon, M., Ed.; Elsevier Scientific Pub. Co.: Amsterdam, 1974, pp. 683-7.
91. Richard, E.P. Jr.; Goss, J.R.; Arntzen, C.J. Weed Sci. 1979, 27, 684-8.
92. Croft, S.M.; Arntzen, C.J.; Vanderhoff, L.N.; Zettinger, C.S. Biochim. Biophys. Acta 1974, 335, 211-7.
93. Shaner, D.L.; Lyon, J.L. Plant Sci. Lett. 1978, 15, 83-7.
94. Kitchen, L.M.; Witt, W.W.; Rieck, C.E. Weed Sci. 1981, 29, 513-6.
95. Kitchen, L.M.; Witt, W.W.; Rieck, C.E. Weed Sci. 1981, 29, In Press.
96. Ali, A.; Fletcher, R.A. Can. J. Bot. 1978, 56, 2196-202.
97. Uotila, M.; Evjen, K.; Iverson, T.-H. Weed Res. 1980, 20, 153-8.
98. Pihakaski, S.; Pihakaski, K. Ann. Bot. 1980, 46, 133-41.
99. Brecke, B.J.; Duke, W.B. Plant Physiol. 1980, 66, 656-9.
100. Fletcher, R.A.; Hildebrand, P.; Akey, W. Weed Sci. 1980, 28, 671-3.
101. Prendeville, G.N.; Warren, G.F. Weed Res. 1977, 17, 251-8.
102. O'Brein, M.C.; Prendeville, G.N. Weed Res. 1979, 19, 331-4.
103. Olorunsogo, O.O.; Bababunmi, E.A.; Bassir, O. FEBS Lett. 1979, 97, 279-82.
104. Olorunsogo, O.O.; Bababunmi, E.A. Toxicol. Lett. 1980, 5(Sp 1), 148.
105. Baur, J.R. Plant Physiol. 1979, 63, 882-6.

RECEIVED September 21, 1981.

11

Determining Causes and Categorizing Types of Growth Inhibition Induced by Herbicides

F. D. HESS

Purdue University, Department of Botany and Plant Pathology, West Lafayette, IN 47907

Sustained plant growth requires cell enlargement and division. Growth can be influenced by an inhibition of one of these processes or by a disruption once initiated. Inhibition and disruption of growth are caused by different mechanisms and result in different effects on plants; therefore, they should be considered separate. Enlargement can be inhibited by metabolic changes decreasing cellular turgor pressure, by decreasing the production, transport, or function of any wall loosening factors (e.g., those induced by IAA), or by an increase in cell wall rigidity prior to enlargement. Disruption of enlargement occurs if bonding between cellulose microfibrils is reduced or if improper microfibril orientation occurs during primary wall synthesis. Compounds that affect plant metabolism can arrest the cell cycle in G_1, S, or G_2, thus these compounds inhibit cell division by inhibiting the onset of mitosis. Disruption of cell division occurs during mitosis and is the result of the mitotic spindle being absent or present but not functional.

When determining the mechanism of action of herbicides affecting growth, the type of growth aberration occurring must be identified. To link observed effects of herbicide action with cell division or enlargement, treatment durations and concentrations inducing aberrations must correlate with those first inhibiting growth. Herbicides acting by inhibiting growth may have diverse biochemical sites of action. Inhibition can result from interference at a single site of action or several sites of action simultaneously. Inhibition of growth may be a primary response of herbicide treatment or can be secondary,

0097-6156/82/0181-0207$05.75/0

in which case growth inhibition may occur late in the sequence of events leading to weed death. For example, compounds that inhibit electron flow in photosynthesis and subsequently induce photooxidation of carotenoids and chlorophylls (e.g., triazines, ureas, and uracils) inhibit growth, but this inhibition is a secondary effect. In other cases inhibition of growth may be the primary cause of weed death. This is true for preemergence herbicides (e.g., dinitroanilines, chloroacetamides, and many carbamates) that inhibit growth shortly after germination, but before weed seedlings emerge from the soil. When their food reserves are depleted through respiration, the seedlings die. This chapter will discuss only those herbicides which inhibit growth as their primary mode of action.

Even though growth inhibition is an important herbicidal effect, there is no uniform terminology to describe the various aberrations induced. Herbicides inducing inhibition of growth have been termed "mitotic poisons" or are claimed to have "meristematic activity", when in fact the inhibition of growth has nothing to do with mitosis or the meristem. Other compounds inhibiting growth are termed "germination inhibitors". A true germination inhibitor, such as methyl bromide, kills the seed and prevents the onset of germination. However, growth inhibitor herbicides disrupt a vital process after germination has been initiated. This chapter attempts to clarify the meaning of growth inhibition induced by herbicides and to categorize the types of growth inhibition. Currently used growth inhibitor herbicides will be categorized as to their mode of inhibition.

Most commonly, growth is considered an irreversible increase in size. A more useful definition for discussing growth inhibitor herbicides would be "a combination of cell division and cell enlargement which leads to an irreversible increase in size". An inhibition of either one of these processes will result in an eventual inhibition of growth.

Several laboratory methods are used to study growth inhibition. Most methods measure root growth at intervals after herbicide treatment. One method consists of pregerminating small seeded broadleaf or grass species and transplanting them to square disposable petri plates containing quartz sand (1) that has been saturated with a nutrient solution [e.g., Hoagland's solution (2)], containing the herbicide. The initial position of the root tip is marked on the petri plate lid. At intervals, the new root position is marked. At the end of the experiment, the increase in root length is plotted with respect to treatment time. Another system of growing plants for inhibition studies is to use "growth pouches". These clear plastic pouches are quite suitable for measuring and recording root growth of small seeded plants (3).

After growth inhibition studies are completed, times and concentrations causing the first significant growth inhibition should be used to further categorize the inhibition. Detected growth inhibition must be the result of an interference with normal cell enlargement, cell division or a combination of both processes (Figure 1).

Cell Enlargement

Interference with normal cell enlargement will result in decreased growth even if cell division is unaffected. To measure growth inhibition caused by a change in cell enlargement, the processes of cell division and cell enlargement must be separated. Oat (*Avena sativa* L.) coleoptiles (4) and pea (*Pisum sativum* L.) hypocotyls (5) are commonly used to study cell enlargement in the absence of cell division. The increase in coleoptile or hypocotyl length is mainly enlargement of previously formed cells (6). Another useful technique in studying cell enlargement is to measure elongation in the alga *Chara* or *Nitella*. A single internode cell will elongate from a few μm to several cm in length (Figure 2).

Aberrant growth occurs after an inhibition or a disruption of the cell enlargement process (Figure 1). These two yield different effects and result from different physiological and biochemical disorders.

Inhibition of Cell Enlargement. In a cell with a wall capable of extension, enlargement occurs when hydrostatic pressure inside the cell exerts a force against the wall. The wall is composed of cellulose microfibrils which slip past one another during enlargement. An inhibition of cell enlargement can be due to membrane damage, metabolic changes within the cell, or changes in processes (e.g., hydrogen ion extrusion) necessary for wall yielding.

Changes in turgor pressure can be caused by herbicides that affect the structure or permeability of the plasma membrane. Several herbicides are reported to disrupt membrane structure [e.g., paraquat (1,1'-dimethyl-4,4'-bipyridinium ion) (7), acifluorfen {5-[2-chloro-4-(trifluoromethyl)phenoxy]-2-nitrobenzoic acid} (8)]. The result of severe membrane disruption is cellular death; therefore, inhibition in cell enlargement is of little importance.

Metabolic disruptions may result in inhibition of cell enlargement. Key (9) found that actinomycin D, an inhibitor of DNA directed RNA synthesis (10), and puromycin, an inhibitor of protein synthesis, will prevent cell enlargement in soybean *Glycine max* L. Merr.) hypocotyls. Key concluded "RNA and protein synthesis are essential for the process of cell elongation to proceed at a normal rate". The following year, Cleland reported the inhibition of cell enlargement caused by actinomycin D was not caused by an inhibition of auxin-induced cell wall loosening

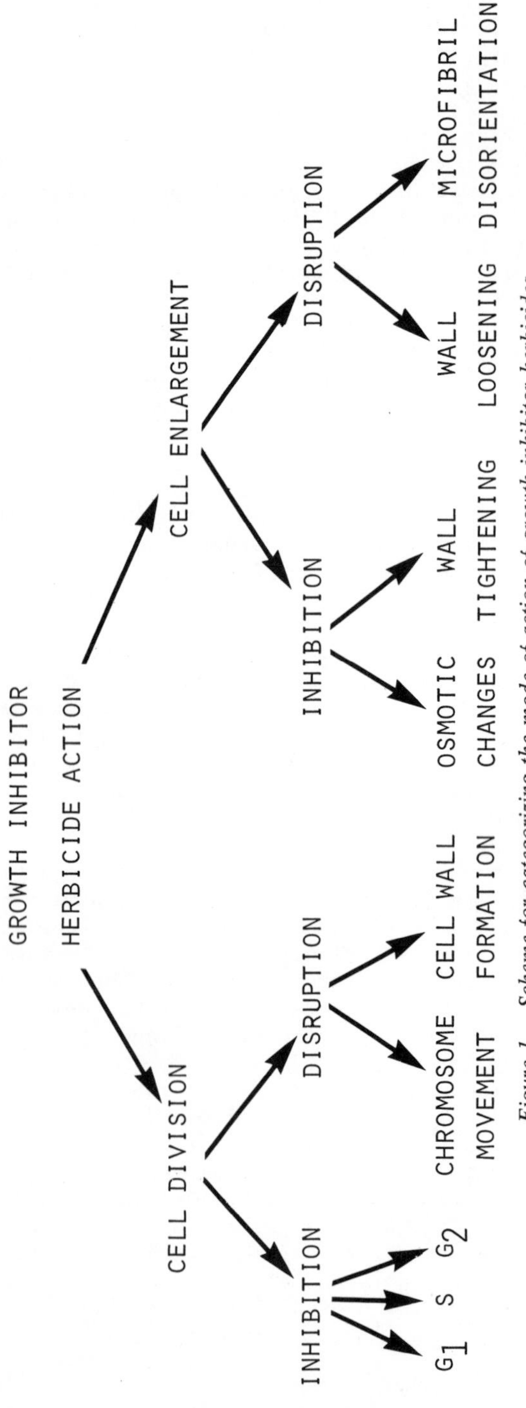

Figure 1. Scheme for categorizing the mode of action of growth inhibitor herbicides.

Figure 2. Internode cell enlargement in the alga Chara. *The two photographs are of the same* Chara *plant. The bottom photograph was recorded 10 days after the top photograph. Magnification = 10×.*

(11), but rather concluded RNA synthesis "is necessary for some other factor required for elongation".

For the wall to yield to osmotic pressure, hydrogen ions are pumped into the wall, which presumably then activate hydrolytic enzymes that digest load bearing polysaccharides (12). This phenomenon is termed the acid growth hypothesis (12, 13). Compounds that interfer with this process impede cell enlargement. The respiration inhibitor CCCP [carbonyl cyanide m-chlorophenylhydrazone], inhibits cell elongation in oat coleoptiles and interferes with hydrogen ion excretion (14). Shimabukuro and Shimabukuro (15) reported that 100 μM diclofop-methyl {methyl 2-[4-(2,4- dichlorophenoxy)-phenoxy]propanoate} reduced the rate of IAA-induced elongation in oat coleoptiles within 50 min and caused complete cessation of elongation within 3 h. Concomitant with the inhibition of elongation was an inhibition of hydrogen ion extrusion in oat coleoptiles treated with 10 μM IAA and 100 μM diclofop-methyl.

Premature cell wall rigidity may also cause cell enlargement inhibition by preventing the microfibrils from slipping past one another in response to turgor pressure. This phenomenon is often termed cell wall tightening. Although, no herbicides are known to inhibit cell enlargement by this mechanism, high calcium concentrations will (16).

Potentially, herbicides can inhibit cell enlargement by interfering with any one of the four cell enlargement factors described by Cleland (12). The first factor is wall extensibility. This process is thought to be under the control of auxin, which induces the excretion of a "wall loosening factor" (perhaps hydrogen ions). Herbicides could inhibit cell enlargement by influencing the production or action of auxin, the production or export of the wall loosening factor, or the ability of the wall to respond to the wall loosening factor. The second factor, namely osmotic concentration, has a major influence on the turgor pressure that must be developed and maintained. Differences in osmotic concentration between the inside and outside of the cell create a significant potential for cell enlargement. A herbicide that interferes with a plant's ability to produce osmotic solutes, to transport them to the enlarging cells, or to move them into the cells, will inhibit cell enlargement. A third factor is the "turgor pressure that must be exceeded" (12) before cell enlargement can occur. Enhanced wall rigidity induced by herbicides could increase the yield turgor to where cell enlargement was inhibited. The final factor affecting cell elongation is the potential for water conductivity into the cell. If a herbicide could modify the cell membrane so that permeability to water is affected, cell enlargement, and hence growth, would be inhibited.

Several herbicides inhibit cell enlargement; however, the mechanism is unknown. In wild oats, diallate [S-(2,3-dichloroallyl)diisopropylthiocarbamate] and triallate [S-(2,3,3-

trichloroallyl)diisopropylthiocarbamate] inhibit shoot growth at concentrations below those that influence mitosis (17). These observations suggest the inhibition of cell enlargement is responsible for the decrease in growth. Wilkinson (18) found that dalapon (2,2-dichloropropionic acid) at 1.0 mM inhibited corn (*Zea mays* L.) coleoptile elongation. Alachlor [2-chloro-2',6'-diethyl-*N*-(methoxymethyl)acetanilide] and metolachlor [2-chloro-*N*-(2-ethyl-6-methylphenyl)-*N*-(2-methoxy-1-methylethyl) acetamide] inhibited cell enlargement in oat coleoptiles at times (24 h) and concentrations (50 μM) that significantly inhibited root growth (19). Cutter, Ashton, and Huffstutter (20) reported bensulide [*O*,*O*-diisopropyl phosphorodithioate *S*-ester with N-(2- mercaptoethyl) benzenesulfonamide] at 50 μM inhibited root growth in oat root tips. At this concentration, cell division was "not completely inhibited"; therefore, they concluded that growth inhibition was caused by an inhibition of cell enlargement. EPTC (*S*-ethyl dipropylthiocarbamate) concentrations that inhibit growth in corn seedlings do not affect cell division, whereas cell enlargement is inhibited (J. D. Holmsen, personal communication).

Disruption of Cell Enlargement. A herbicidal effect can disrupt rather than inhibit normal enlargement. Because growth effects resulting from an inhibition or a disruption of enlargement are so different, they should be considered separate.

IAA will disrupt cell enlargement by inducing wall loosening (21). A herbicide known to disrupt cell enlargement is 2,4-D [(2,4-dichlorophenoxy)acetic acid]. Expansive growth in soybean hypocotyls increased by 66% during the first 2 h, 71% during the second 2 h, and 100% between 4 and 8 h after treatment with 45 μM 2,4-D (9).

Enlargement of individual root cells is essentially unidirectional (longitudinally anisotropic). The other disruption herbicides can have on root cell enlargement is to change the order of cell enlargement from cylindrical (anisotropic) to spherical (isotropic). Anisotropic cell enlargement is influenced by the orientation of cellulose microfibrils in the primary wall adjacent to the plasma membrane. The microfibrils are oriented in a transverse manner around the longitudinal walls (e.g., 22, 23). This orientation gives preferential strength in the transverse direction, thus when turgor pressure occurs in the cell, the wall tends to yield (enlarge) in the longitudinal direction. This process is termed the multinet theory of cell enlargement (24). Compounds that disrupt the microfibril orientation will disrupt anisotropic enlargement of cells. The most studied compound of this type is the drug colchicine. Green *et al*., have shown colchicine treatment is associated with the disruption of normal microfibril orientation in the cell wall. As a result, isotropic cell enlargement occurs in *Nitella* (25). Colchicine inhibits (*in vivo*) formation of microtubules (26), which are theorized to orient the cellulose microfibrils during primary wall synthesis (27). Trifluralin (α,α,α-trifluoro-2,6-dinitro-*N*,*N*-dipropyl-*p*-toluidine) also affects cell enlargement (28). Treated root cells in the zone of elongation enlarge isotropically, (Figure 3) rather than anisotropically (Figure 4). Because trifluralin also disrupts cell division, the zone of elongation in treated roots becomes close to the tip (Figure 3). Isotropic cell enlargement also occurs when *Chara* is treated with trifluralin during cell enlargement. Internodes that are small at the beginning of treatment (insert in Figure 2) will enlarge isotropically (Figure 5) rather than anisotropically (enlarged internode in Figure 2). The result of isotropic cell enlargement is a root tip with the characteristic "club shape" (Figure 6). Because of the proposed involvement of microtubules in ordered cell enlargement, swollen root tips suggest the herbicide is influencing microtubule structure or function.

Cell Division

The cell division cycle as described by Howard and Pelc (29) consists of G_1, S, G_2, and M. G_1 and G_2 are termed Gap 1 (pre-DNA synthesis period) and Gap 2 (pre-mitotic period). The replication of DNA occurs during S (synthesis). The stages of G_1, S, and G_2 are termed interphase. Mitosis (M), which occurs over a short time in comparison to G_1, S, and G_2, is the most visible portion of the cell cycle.

During mitosis, the mitotic spindle is responsible for the physical movement of chromosomes. The spindle is composed of hundreds of microtubules. During prophase, numerous microtubules form at the kinetochores and then move the chromosomes first to a central plane (metaphase) and then to the poles during anaphase. At telophase, microtubules are present at the site of cell plate formation. At interphase, microtubules are located in the cytoplasm near the cell wall. Microtubules do not move from one location to another during the cell cycle (e.g., from the cell wall at interphase to the kinetochore of the chromosome during prophase), but rather are depolymerized at one location into a common pool of protein subunits called tubulin. At their new location, the microtubules repolymerize from the tubulin pool.

Methods Used to Study Cell Division. Three methods are commonly used to study cell division. One method which studies cell population increases in single cell algae (30) will not discriminate between types of cell division aberrations, but is the easiest system to use. The test herbicide is added to a cell population (10^4 to 10^5 cells per ml) and incubated for 24 or 48 h. The cell population is then assayed with a Coulter counter, spectrophotometer (turbidity), or hemocytometer. This

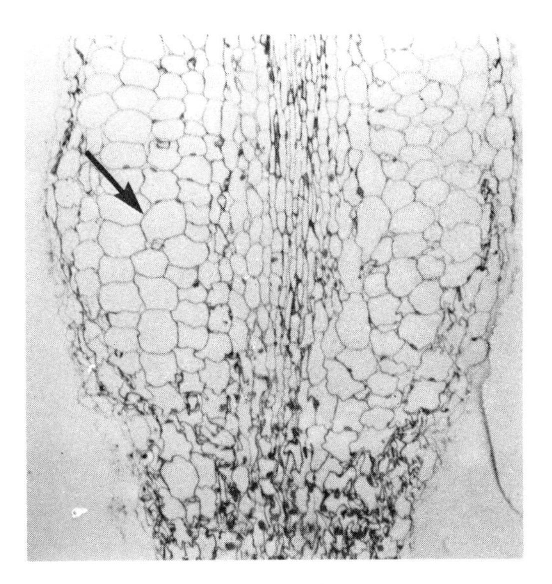

Figure 3. Lateral root tip from a pea plant. The plant was grown in an aqueous saturated solution of trifluralin for 48 h. Note the isotropic cells in the zone of elongation (arrow). Magnification = 70×.

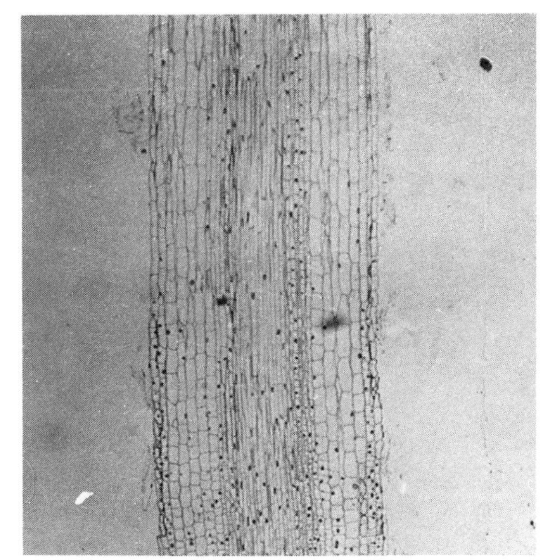

Figure 4. Zone of elongation in lateral root from a nontreated pea plant. Characteristic anisotropic cells are present in the zone of elongation. Magnification = 70×.

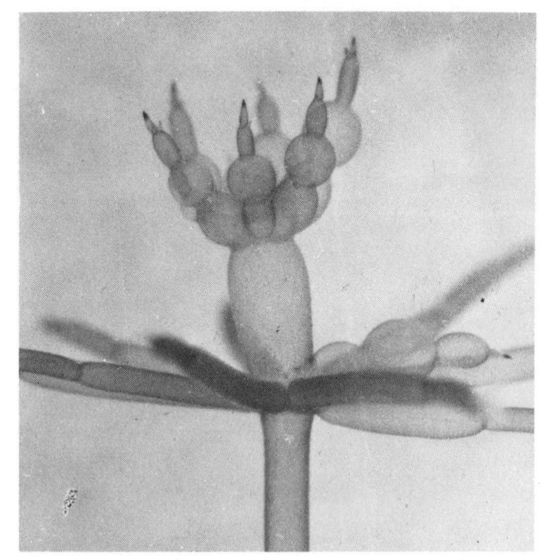

Figure 5. Apex of Chara *plant treated with an aqueous saturated solution of trifluralin for 9 days. Isotropic cell enlargement has occurred in the internode cell and in lateral cells arising from the node of the main axis. Magnification = 12×.*

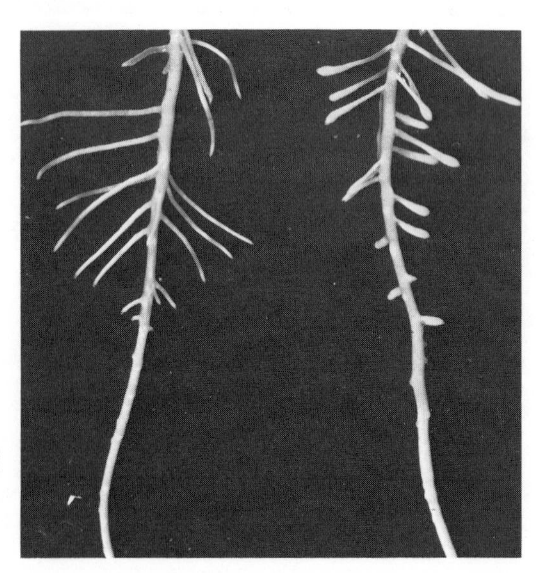

Figure 6. Lateral root development in peas. Left is nontreated. Right is treated with trifluralin. Note the typical "club shape" appearance of treated lateral root tips.

assay is not specific for cell division because photosynthesis inhibition, membrane disruption, and cell enlargement inhibition will also inhibit cell population increases. The second method is to fix, dehydrate, and embed herbicide-treated tissues in paraffin. Detailed methodology for this procedure can be found in Jensen (31) or Sass (32). Tissues from longitudinal sections are then evaluated by light microscopy for frequency of mitotic figures and identification of aberrant mitotic stages. The third method of evaluating herbicidal effects on the cell cycle is to prepare a "root tip squash". Details of this procedure are described by Van't Hof (33). Once the tissue has been prepared, a detailed cell cycle kinetic analysis can be conducted. The procedure involves categorizing cells as being in interphase, prophase, metaphase, anaphase plus telophase, or as aberrant. This analysis will determine if the herbicidal effect on cell division is at interphase (G_1, S, G_2) or mitosis.

Caution is warranted when collecting cell cycle kinetic data if the effect is on mitosis. If cells contain numerous chromosomes, an accumulation of cells with chromosomes in the blocked metaphase configuration (clumped chromosomes) may be interpreted as an accumulation of prophase division figures. If the chromosomes are not moved to the metaphase plate by the spindle apparatus, the chromosome arrangement will remain random and the cells will appear to be in prophase. The chromosomes, however, will have become thickened and uncoiled as if in metaphase. These division figures should not be considered in normal prophase, but rather should be listed as "aberrant" or as in a "blocked metaphase" condition. A test species should be selected that has an adequately low chromosome number to allow a clear separation of chromosome shape and size between prophase and metaphase.

Regardless of the procedure that is selected, variables of treatment time and herbicide concentration should be studied. The shortest treatment time and the lowest herbicide concentration that induced significant inhibition of growth should be used in the mitotic analysis. In addition, the same plant species and preferably the same tissue type used for the growth inhibition study should be used for the cell cycle analysis. High concentrations may induce effects on cell division differing from those affecting growth at lower concentrations. For example, Banting (17) reported diallate treatment of oats at 15 μM resulted in a significant decrease in shoot growth, whereas the effect on mitosis was minor. At 60 μM diallate there was a significant disruption of mitosis. At long treatment times (24 to 48 h), the effect on cell division may be a secondary expression of metabolic disruption that occurred immediately after treatment. Observing cell division at long treatment durations will aid in understanding whether cell enlargement or division is causing the inhibition of growth, but will not reveal if the effect is primary or secondary.

Types of Cell Division Effects. Classification of herbicidal effects on cell division is not uniform. This has lead to confusion about the action of herbicides on cell division. Terms such as "mitotic poisons", "meristem active", and "mitotic inhibitors" have been used to describe the same effect of a herbicide on cell division. A more useful classification of herbicidal effects would be to divide herbicides into 2 classes: those inhibiting cell division and those disrupting cell division (Figure 1). Inhibition of cell division will result in treated meristems that only contain interphase cells. If cell division is disrupted, one or more mitotic stages normally present in the meristem tissue will not be found. These two effects on cell division result from different mechanisms.

Cell Division Inhibition. If the mitotic index (% of cells in mitosis) decreases to zero during the first few hours of treatment (approximately 8 h), the compound is inhibiting some metabolic process in interphase (G_1, S, or G_2). The location and possible cause of the inhibition, which can be determined by autoradiographic techniques, involves treating root meristems with radioactive thymidine after herbicide treatment. The details of this procedure have been published by Van't Hof (33). An alternative to the above procedure is to treat root tissue with herbicide plus radioactive thymidine or uridine and follow the increase of radioactivity precipitable by trichloroacetic acid (34).

Disruption of key metabolic processes, such as DNA, RNA, and protein synthesis in the meristematic area will cause a cell cycle arrest in G_1, S, or G_2. A review of this subject has been published by Rost (35). Hydroxyurea has been reported to inhibit cell division by inhibiting DNA synthesis without influencing RNA or protein synthesis (36). Cell division was not stopped immediately by hydroxyurea, but rather mitosis ceased to occur only after a period of time equivalent to G_2, suggesting those cells which had completed DNA synthesis (S) proceeded to and through mitosis. RNA synthesis inhibition and subsequent cell cycle arrest in G_1 and G_2 occur with actinomycin D treatment (37). When actinomycin D was added to proliferative (those growing in the presence of sucrose and dividing normally) excised pea root tips, the progression of G_1 to S was inhibited after an 8 h lag period. Cycloheximide, a well known inhibitor of protein synthesis, has been shown to block cell cycle progression from G_1 to S and G_2 to M (37). If proliferative cells were treated with the protein synthesis inhibitor puromycin, radioactive leucine incorporation was inhibited only 9% after a 4 h treatment, yet mitosis was inhibited over 50% (37).

Herbicides have been reported to inhibit cell division. Alachlor and metolachlor at 1 μM significantly reduced the frequency of cell division in oat root tips after 30 h (19). CDAA (N-N-diallyl-2-chloroacetamide) was reported to reduce mitosis in barley (*Hordeum vulgare* L.) roots nearly 90% after 96 h at 57 μM (38). Propachlor (2-chloro-N-isopropylacetanilide) totally inhibited mitosis in onion (*Allium cepa* L.) root tips after an 18 h treatment with 75 μM (39). At 20 μM cell division was reduced approximately 50% and cell enlargement was reduced 40% in oat coleoptiles (39). After 24 h, 100 μM ioxynil (4-hydroxy-3,5-diiodobenzonitrile) reduced the mitotic index in broad bean (*Vicia faba* L.) and pea root tips (40). Few herbicides that inhibit cell division have been studied in adequate detail to locate the site of the block. A notable exception is the herbicide chlorsulfuron (DPX 4189, 2-chloro-N-{[(4-methoxy-6-methyl-1,3,5-triazin-2-yl)amino]carbonyl}-benzenesulfonamide) (41). Ray reported a 50% reduction in corn growth 3 h after treatment with 28 μM chlorsulfuron. Mitosis in broad bean root tips was significantly reduced by 2.8 μM, whereas in three different tests, cell enlargement was not influenced with concentrations of 28 μM. Thymidine incorporation into DNA was inhibited in corn root tips after a 1 h treatment with 2.8 μM. After 6 h, the inhibition was 80 to 90%. At this time and concentration (6 h, 2.8 μM), protein and RNA synthesis were not affected. These results suggest that chlorsulfuron is inhibiting a specific process in G_1, S, or G_2.

Cell Division Disruption. If a cell cycle kinetic analysis reveals the herbicide effect is not caused by inhibition, the effect may be on mitosis. In this instance, the kinetic analysis will identify mitotic stages not recognizable as prophase, metaphase, anaphase, or telophase.

Abnormal chromosome configurations are often caused by an effect on the spindle apparatus. This effect is attributed to an abnormal function or an absence of the spindle. Herbicides are known to induce both types of effects. Observations with the light microscope may not distinguish between the two possibilities. In this case, the electron microscope can be used to assess the presence or absence of microtubules.

Colchicine is the classic compound associated with the absence of microtubules in plant cells (26). This drug binds to tubulin and thereby prevents microtubule polymerization of the tubulin pool within the cell (42). The herbicide trifluralin (43, 44, 45) induces microtubule absence in plant tissue (Figure 7). Trifluralin-tubulin binding reactions have not been possible with higher plant tubulin because higher plant tubulin has not been isolated and purified to the degree necessary for drug binding reactions. However, trifluralin binds to flagellar tubulin isolated from the alga *Chlamydomonas* (46) and prevents flagellar regeneration (47). Trifluralin does not bind to

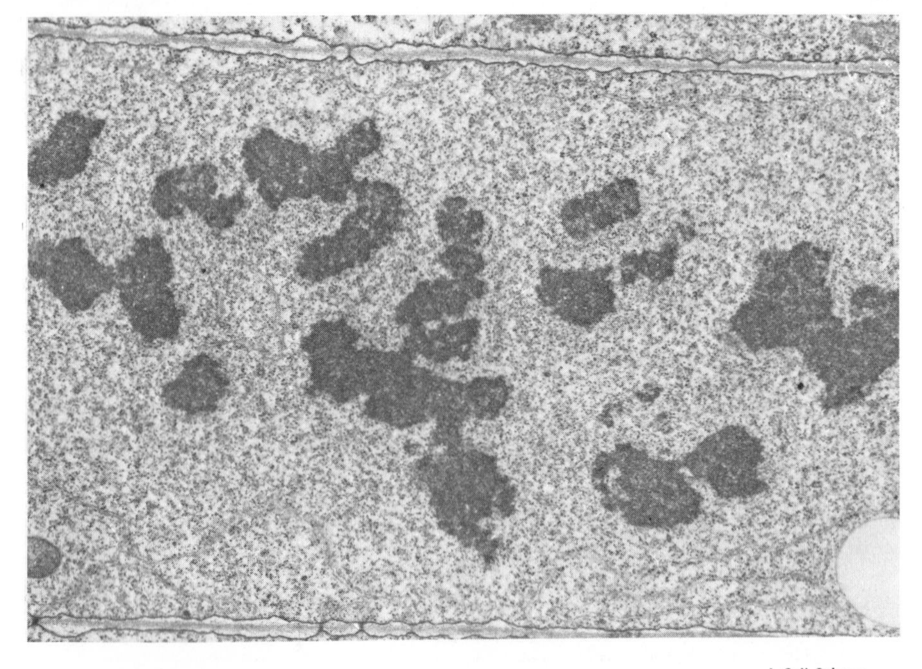

J. Cell Science

Figure 7. Prophase division figure in a cotton lateral root treated with an aqueous saturated solution of trifluralin for 2 h (45). Microtubules are absent. Magnification = 8,000×.

tubulin isolated from animal tissue nor prevent in vitro polymerization of this tubulin into microtubules (44). This result is consistent with the observation that trifluralin does not affect cell division in animal tissue (46). Similarly, there are significantly different colchicine sensitivities between animal and plant tissues (27).

When binding and in vitro polymerization reactions can be conducted with higher plant tubulin, one will be able to ascertain if trifluralin prevents microtubule polymerization by binding to tubulin. Recently, trifluralin and other dinitoaniline herbicides have been shown to inhibit calcium uptake by mitochondria (48). This results in cytoplasmic concentrations of calcium that are claimed to prevent microtubule polymerization. Oryzalin [3,5-dinitro-N^4, N^4-dipropylsulfanilamide]), a close analog of trifluralin, inhibits mitochondrial calcium uptake "half-maximal" at 20 μM, whereas at concentrations below 10 μM there is no effect on calcium uptake (48). Oryzalin will disrupt cell division 50% at 1.25 μM and nearly 100% at 2.5 μM. Regardless of the mechanism involved in inhibiting microtubule formation, the disruption of cell division will eventually lead to an inhibition of growth.

Another mechanism of mitotic disruption, the abnormal function of spindle microtubules, has been reported for the carbamate herbicides propham (isopropyl carbanilate) and chlorpropham (isopropyl m-chlorocarbanilate). Many researchers have reported propham and chlorpropham disrupt mitosis (e.g. 39, 49, 50). The cause of the disruption was unknown until Hepler and Jackson studied the effect of propham on dividing liquid endosperm cells of African blood lily (Haemathus katherinae Baker) with the electron microscope (51). They reported propham treatment (56 μM) for 2 h caused microtubules to lose their "parallel alignment and become oriented in radial arrays" (Figure 8). The microtubule orientations were not random, but rather were aligned in a "multipole spindle apparatus". Because of the multipole spindle, several micronuclei, which coalesced into a single large polyploid nucleus, were produced (52). In 1974, Coss and Pickett-Heaps published (53) evidence suggesting propham's site of action was the microtubule organizing center (MTOC). Improper microtubule function effectively disrupts cell division and can account for the reported inhibition of growth. At high concentrations (>0.5 mM) propham also inhibits RNA, DNA, and protein synthesis (54) as well as prevents microtubule formation (53).

Compounds associated with a microtubule absence during mitosis often reduce microtubule frequency when studied at short time intervals (0.5 h) or at low concentrations. For example, treatment of cells with low concentrations of colchicine (55) or trifluralin (43, 45) results in a partial loss of microtubules. The remaining microtubules are sometimes disoriented (45).

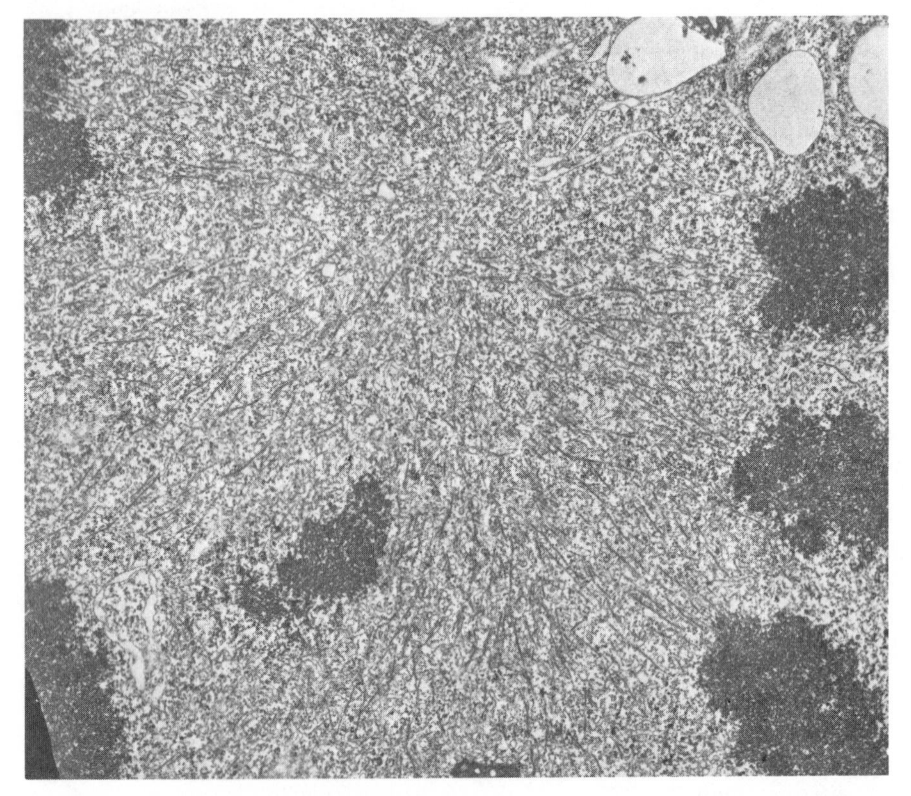

J. Cell Biology

Figure 8. Multipole spindle apparatus (51). Microtubules are present but have functioned abnormally. Haemanthus katherinae *liquid endosperm cells had been treated with 56 μM propham. Magnification = 10,000×.*

Other herbicides are reported to disrupt mitosis; however, in most instances the mechanism of disruption is unknown. DCPA (dimethyl tetrachloroterephthalate) at 30 μM was reported to interfere with nuclear division in corn roots (56). Diallate treatment (60 μM) induced the formation of abnormal division figures and reduced metaphase, anaphase, and telophase in wild oat (*Avena fatua* L.) stem tissue (17). Prasad and Blackman, who found (57) 1 mM dalapon induced abnormal cell division in corn root tips, concluded mitosis was arrested at prophase. Carlson *et al.* (58) reported pronamide [3,5-dichloro(*N*-1,1-dimethyl-2-propynyl)benzamide] induced the occurrence of "arrested metaphase" in oat root seedlings within 0.5 h after treatment with 20 μM. Bartels and Hilton (44) found pronamide (100 μM) caused a "loss of both cortical and spindle microtubules in root cells".

Two interesting observations are frequently reported for herbicides that disrupt mitosis. First, tissue differentiation and maturation continues after mitotic disruption. As a result, differentiated cells occur abnormally close to the root tip (e.g. 20, 50, 56, 59, 60). Second, compounds that disrupt mitosis often induce abnormal stem formation (e.g. 61-65). The xylem elements become shortened, twisted, and discontinuous. This abnormality results in swollen hypocotyls that are often brittle.

Compounds may not affect chromosome movement, but can disrupt or prevent cell plate formation at telophase. This effect is classified as a cell division disruption rather than an inhibition because the early stages of mitosis occur and appear normal. The cell division sequence is not disrupted until cell plate formation at telophase. Although no herbicides are known to disrupt or inhibit cell wall formation as their sole mode of action, the compound caffeine (1,3,7-trimethylxanthine) is a specific inhibitor of cell plate formation at telophase (35). Paul and Goff reported (66) that aminophylline, theophylline (1,3-dimethylxanthine), caffeic acid, and caffeine blocked cell wall formation at telophase by inhibiting vesicle fusion.

The Occurrence of Multinucleate Cells

Multinuclei observed by light microscopy are almost universally reported to occur in cells treated with mitotic disrupting herbicides (e.g. 67, 68, 69). Authors suggest that nuclear division without cell plate formation is an injury response of mitotic disrupting herbicides. They often report the absence of normal division figures and the presence of clumped chromosome arrangements. Observation of published micrographs reveals that most of the reported multinuclei are in proximity. Observation of paraffin sections, with an average thickness of 6 to 10 μm, may not resolve thin connections between the spheres; therefore, uninucleate cells may be interpreted as being multinucleate. Studies of thin sections (1 μm thick) often reveal small connections between the proximally located spherical nuclear lobes (Figure 9). Walne (70) recognized that some of the multinucleate cells induced by colchicine could be a misinterpretation of polymorphic uninuclei. With staining techniques, Walne found small interconnections between many nuclei of Chlamydomonas eugametos cells. In an electron microscope study, Pickett-Heaps (26) reported the occurrence of polymorphic nuclei after colchicine treatment. In root tip tissue treated with trifluralin, polymorphic nuclei have been observed by electron microscopy (Figure 10, 45).

There are three explanations for multinucleate cells. First, one or more blocked metaphase chromosomes may become separated from the main group (Figure 11). When the nuclear membrane forms, several nuclei could result. This occurs in colchicine treated tissue (71). Second, the meristem contains cells in all stages of the cell cycle. If treatment induces microtubule absence in cells at late anaphase or early telophase, nuclear envelope reformation will result in a binucleate cell. Cells containing two nuclei are at times termed "multinucleate". Third, polymorphic nuclei may have very thin connections between lobes (Figures 9 and 10). These cells may appear as multinucleate when in fact they are uninucleate. To suggest a mitotic disrupting herbicide has induced a "multinucleate condition", analysis must prove more than two discrete nuclei are actually present in the cell.

Conclusion

As shown in Figure 1, inhibition of growth can be caused by numerous aberrant conditions within the growth regions of plants. If characterization beyond "growth inhibition" is attempted for a given herbicide, the various possibilities must be thoroughly investigated. To learn if the effect is on cell division, cell enlargement, or both, experiments must be conducted at times and concentrations where growth inhibition first occurs. Then one must ascertain whether the effect is caused by an inhibition or a disruption of the process. After this is determined, evaluation of the aberrancy may reveal the precise site of action. Growth inhibition should not be described beyond what is actually known about the cause.

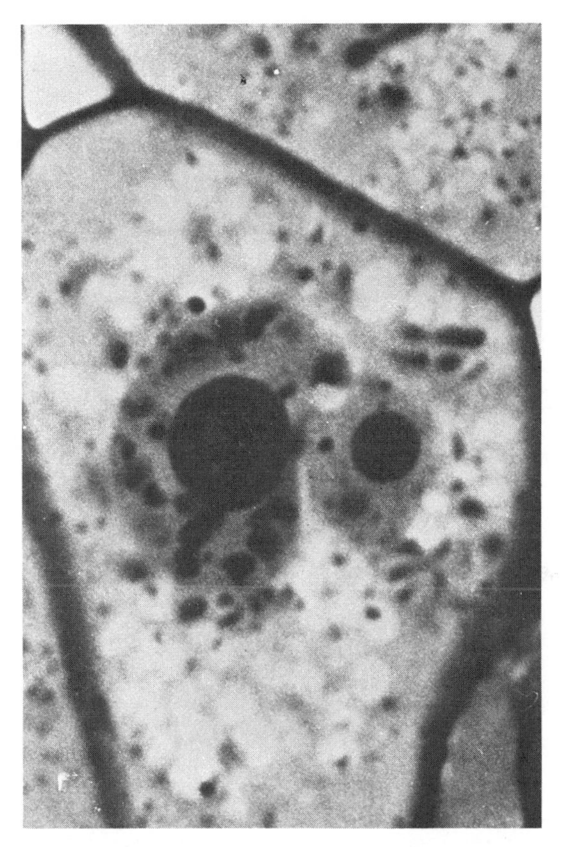

Figure 9. Section (1.5 μm thick) from a cotton lateral root tip treated with trifluralin for 24 h. Note the presence of a polymorphic nucleus. Magnification = 3,000×.

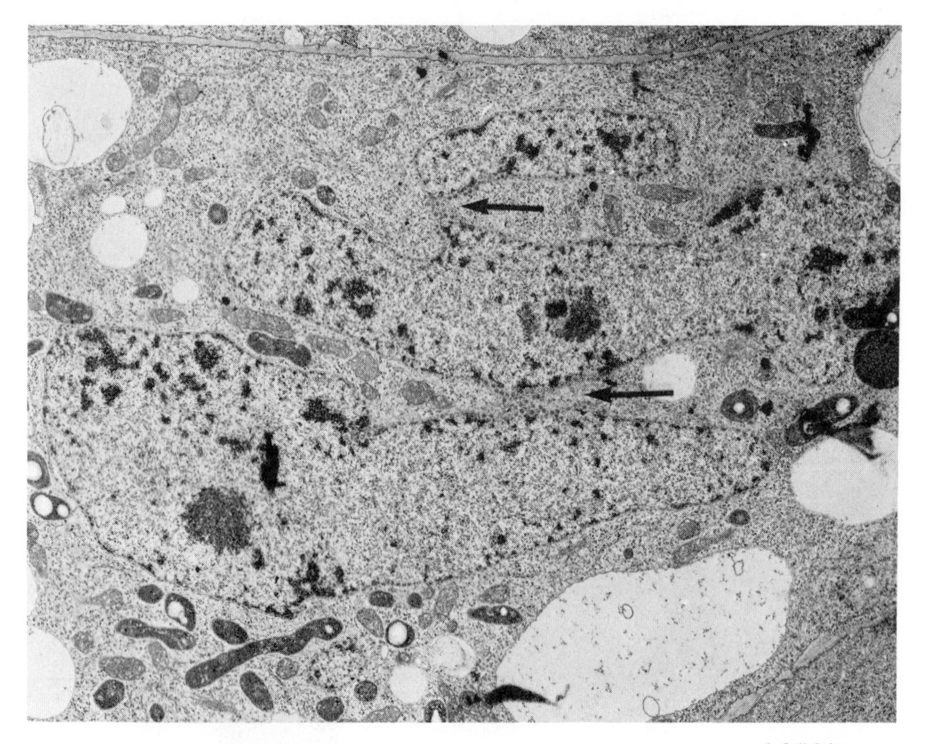

J. Cell Science

Figure 10. Electron micrograph of a polymorphic nucleus (45). There is a thin connection between each nuclear lobe (arrows). The micrograph is from cotton lateral root tissue treated with trifluralin for 24 h. Magnification = 5,500×.

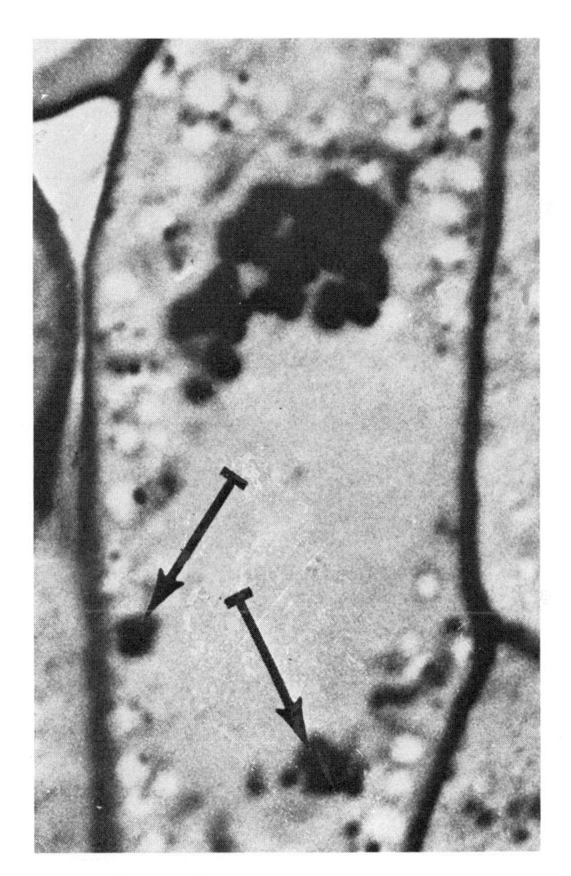

Figure 11. Aberrant division figure from a lateral root of cotton treated with an aqueous saturated solution of trifluralin for 24 h. Some chromosomes (arrows) are widely separated from the main group. Magnification = 3,000×.

Literature Cited

1. Parker, C. Weeds 1966, 14, 117-21.
2. Hoagland, D. R.; Arnon, D. I. Univ. California Agric. Exp. Stn. Circ. 347 1950, p 32.
3. Porter, F. E.; Nelson, I. S.; Wold, E. K. Crops & Soils 1966, 18, No 8, 10-2.
4. Bonner, J. J. Gen. Physiol. 1933, 17, 63-76.
5. Thimann, K. V.; Schneider, C. L. Amer. J. Bot. 1939, 26, 792-7.
6. Nitsch, J. P.; Nitsch, C. Plant Physiol. 1956, 31, 94-111.
7. Prendeville, G. N.; Warren, G. F. Weed Res. 1977, 17, 251-8.
8. Orr, G. L.; Hess, F. D. Pest. Biochem. Physiol. 1981, In Press.
9. Key, J. L. Plant Physiol. 1964, 39, 365-70.
10. Hurwitz, J.; Furth, J. J.; Malamy, M.; Alexander, M. Proc. Nat. Acad. Sci. USA 1962, 48, 1222-30.
11. Cleland, R. Plant Physiol. 1965, 40, 595-600.
12. Cleland, R. Symp. Soc. Exp. Biol. 1977, 31, 101-15.
13. Rayle, D. L.; Cleland, R. E. Plant Physiol. 1970, 46, 250-3.
14. Cleland, R. Proc. Nat. Acad. Sci. USA 1973, 70, 3092-3.
15. Shimabukuro, M. A.; Shimabukuro, R. H. Weed Sci. Soc. Amer. Abstr. 1981, 111.
16. Tagawa, T.; Bonner, J. Plant Physiol. 1957, 32, 207-12.
17. Banting, J. D. Weed Sci. 1970, 18, 80-4.
18. Wilkinson, R. E. Weeds 1962, 10, 275-80.
19. Deal, L. M.; Hess, F. D. Weed Sci. 1980, 28, 168-75.
20. Cutter, E. G.; Ashton, F. M.; Huffstutter, D. Weed Res. 1968, 8, 346-52.
21. Cleland, R. Physiol. Plantarum. 1958, 11, 559-609.
22. Setterfield, G.; Bayley, S. T. J. Biophys. & Biochem. Cytol. 1958, 4, 377-82.
23. Northcote, D. H. Proc. R. Soc. Lond. ser. B, Biol Sci. 1969, 173, 21-30.
24. Roelofsen, P. A.; Houwink, A. L. Acta Bot. Neerl. 1953, 2, 218-25.
25. Green, P. B.; Erickson, R. O.; Richmond, P. A. Ann. N. Y. Acad. Sci. 1970, 175, 712-31.
26. Pickett-Heaps, J. D. Dev. Biol. 1967, 15, 206-36.
27. Hepler, P. K.; Palevitz, B. A. Annu. Rev. Plant Physiol. 1974, 25, 309-62.
28. Hess, F. D. "The mechanism of action of trifluralin: influence on cell division, cell enlargement, and microtubule protein"; Ph.D. Dissertation: Univ. Calif., Davis, 1975; p 194.
29. Howard, A.; Pelc, S. R. Heredity 1953, 6(Suppl.), 261-73.

30. Hess, F. D. Weed Sci. 1980, 28, 515-20.
31. Jensen, William A. "Botanical Histochemistry"; W. H. Freeman & Company: San Francisco, California, 1962; p 408.
32. Sass, John E. "Botanical Microtechniques"; The Iowa State Univ. Press: Ames Iowa, 1958; p 228.
33. Van't Hof, J. "Methods in Cell Physiology", Vol III; Prescott, D., Ed.; Academic Press: New York, 1968; pp 95-117.
34. Rost, T. L.; Bayer, D. E. Weed Sci. 1976, 24, 81-7.
35. Rost, T.L. "Mechanisms & Control of Cell Division"; Rost, T. L.; Gifford, E. M., Eds.; Dowden, Hutchinson & Ross, Inc: Stroudsburg, Pennsylvania, 1977; pp 111-43.
36. Barlow, P. W. Planta 1969, 88, 215-23.
37. Webster, P. L.; Van't Hof, J. Amer. J. Bot. 1970, 57, 130-9.
38. Canvin, D. T.; Friesen, G. Weeds 1959, 7, 153-6.
39. Dhillon, N. S.; Anderson, J. L. Weed Res. 1972, 12, 182-9.
40. Rost, T. L.; Morrison, S. L.; Sachs; E. S. Amer. J. Bot. 1977, 64, 780-5.
41. Ray, T. B. Proc. British Crop Prot. Conf. - Weeds 1980, 1, 7-14.
42. Wilson, L. Life Sci. 1975, 17, 303-10.
43. Jackson, W. T.; Stetler, D. A. Can. J. Bot. 1973, 51, 1513-8.
44. Bartels, P. G.; Hilton, J. L. Pest Biochem. Physiol. 1973, 3, 462-72.
45. Hess, F. D.; Bayer, D. E. J. Cell Sci. 1974, 15, 429-41.
46. Hess, F. D.; Bayer, D. E. J. Cell Sci. 1977, 24, 351-60.
47. Hess, F. D. Exp. Cell Res. 1979, 119, 99-109.
48. Hertel, C.; Quader, H.; Robinson, D. G.; Marme, D. Planta 1980, 149, 336-40.
49. Ennis, W. B. Amer. J. Bot. 1948, 35, 15-21.
50. Ennis, W. B. Amer. J. Bot. 1949, 36, 823.
51. Hepler, P. K.; Jackson, W. T. J. Cell Sci. 1969, 5, 727-43.
52. Jackson, W. T. J. Cell Sci. 1969, 5, 745-55.
53. Coss, R. A.; Pickett-Heaps, J. D. J. Cell Biol. 1974, 63, 84-98.
54. Rost, T. L.; Bayer, D. E. Weed Sci. 1976, 24, 81-7.
55. Jokelainen, P. I. J. Cell Biol. 1968, 39, 68a-9a.
56. Bingham, S. W. Weed Sci. 1968, 16, 449-52.
57. Prasad, R.; Blackman, G. E. J. Exp. Bot. 1964, 15, 48-66.
58. Carlson, W. C.; Lignowski, E. M.; Hopen, H. J. Weed Sci. 1975, 23, 155-61.
59. Peterson, R. L.; Smith, L. W. Weed Res. 1971, 11, 84-7.
60. Scott, M. A.; Struckmeyer, B. E. Bot. Gaz. 1955, 117, 37-45.
61. Kirby, C. J.; Standifer, L. C.; Normand, W. C. Proc. Southern Weed Conf. 1968, 21, 343.
62. Nishimoto, R. K.; Warren, G. F. Weed Sci. 1971, 19, 343-6.
63. Andersen, J. L.; Shaybany, B. Weed Sci. 1972, 20, 434-9.

64. Hess, B.; Bayer, D.; Ashton, F. Weed Sci. Soc. Amer. Abstr. 1974, 17.
65. Struckmeyer, B. E.; Binning, L. K.; Harvey, R. G. Weed Sci. 1976, 24, 366-9.
66. Paul, D. C.; Goff, C. W. Exp. Cell Res. 1973, 78, 399-413.
67. Talbert, R. E. Proc. Southern Weed Conf. 1965, 18, 652.
68. Hacskaylo, J.; Amato, V. A. Weed Sci. 1968, 16, 513-5.
69. Lignowski, E. M.; Scott, E. G. Weed Sci. 1972, 20, 267-70.
70. Walne, P. L. Amer. J. Bot. 1966, 53, 908-16.
71. Hindsmarsh, M. M. Proc. Linnean Soc. N. S. Wales 1953, 77, 300-6.

RECEIVED September 28, 1981.

12

Modes of Herbicide Action as Determined with *Chlamydomonas reinhardii* and Coulter Counting

CARL FEDTKE

Pflanzenschutz, Biologische Forschung, Bayer AG 5090 Leverkusen, Federal Republic of Germany

Synchronously grown Chlamydomonas reinhardii cells were used for the differentiation between different physiological modes of herbicide action. The cells were counted with a Coulter Counter and the volume spectrum was obtained by an adapted Channelyzer. The action of the tested herbicides was classified as follows: (a) growth inhibition in the light without rapid phytotoxicity (metribuzin), (b) bleaching of the cells without rapid phytotoxicity (fluridone), (c) rapid phytotoxicity in the light (nitrofen, paraquat), (d) rapid phytotoxicity in the dark (PCP), (e) direct and rapid inhibition of cell division but not of cell growth (amiprophos-methyl, chlorpropham, trifluralin) and (f) indirect (delayed) inhibition of cell division and growth (alachlor).

Depending on concentration, herbicides may inhibit many different plant metabolic pathways in vitro as well as in vivo. However, often the relevance of an inhibitory activity in vitro for the action on the intact growing plant is not clear. Trifluralin and chlorpropham e.g. affect chloroplast and mitochondrial activities in vitro (1). In the field, however, they inhibit the growth of seedlings and provoke characteristic morphological responses which can not be traced to the effects on the organelles mentioned above. A simple test system more closely reflecting the primary sensitive pathway under field conditions would therefore clearly be helpful.

Unicellular algae have repeatedly been used for the estimation and classification of herbicide actions (2, 3, 4). Chlamydomonas reinhardii appeared to be most suitable since this species proved to be sensitive towards herbicides of many different groups. Chlamydomonas reinhardii strain 11 - 32 a/89 was obtained from the "Algensammlung Göttingen" and was grown on Tris-Acetate-Phosphate medium (5). The culture vessels (250 ml beakers containing 150 ml growth medium) were kept on a rotary shaker at 100 rpm in a 12/12 h light/dark regime. The illumination was 5 000 lx from Osram HQI bulbs and the temperature was 24 C during the light and 18 C during the dark phase. The maximum cell density at which the cultures were diluted was 7×10^5 cells/ml. Under these conditions, the multiplication factor was ca. 4 to 5 in 24 h. The herbicides were dissolved in methanol. Only chemically pure compounds (>99 %) were used.

The cells were counted by a Coulter Counter Model ZB equipped with a 50 µm aperture after appropriate dilution with growth medium. Simultaneously the size distribution was obtained by a Coulter Channelyzer Model C-1000. The sizing is linear with volume and was set to span the range from 70 to 900 μm^3 (Figures 2, 3, etc.).

Cell Cycle Analysis with the Coulter Counter

In the regular cell cycle of *Chlamydomonas reinhardii*, cells grow in the light and divide in the dark (Figure 1). At the beginning of the dark phase, the cells shed their flagella and swell slightly to gain a spherical form. During the dark phase, nuclear, chloroplast, and cell divisions take place, and the daughter cells are liberated. In contrast to other strains, in these experiments the daughter cells were liberated throughout the dark phase and not synchronously at its end (6).

Inoculation of new cultures and Coulter Counter measurements were usually started at the beginning of the dark phase. At this time, fully grown cells are present (Figure 2, trace 0). Five hours later (trace 5), most of the cells have swollen and entered division. Eight hours into the dark phase (trace 8), many mother cells have already liberated the daughter cells, whereas at the end of the dark phase (trace 12) a homogenous population of small, young cells is obtained. These cells grow during the next 12 h light phase to attain the fully grown size again (Figure 3).

The herbicides selected for the present study are given in Table I together with their respective chemical names. Table II presents the activity data towards *Chlamydomonas reinhardii* for these herbicides and also indicates the metabolic pathway or process which is presumably primarily affected by the action of the herbicide and whose inhibition or alteration most likely causes the observed effects. The flagella regeneration represents a process which depends on microtubule polymerization, and which is, therefore, seen as a test for some types of antimitotic activity (7). However, flagellar regeneration is an energy requiring process and is, therefore, inhibited also by the respiratory uncoupler PCP. Interestingly, nitrofen is also active in this test. Summarizing, two broad groups may be formed from the selected herbicides, one interfering with photosynthetic processes and a second interfering with cell division, either directly or indirectly (1, 8, 9). These two groups will be discussed separately.

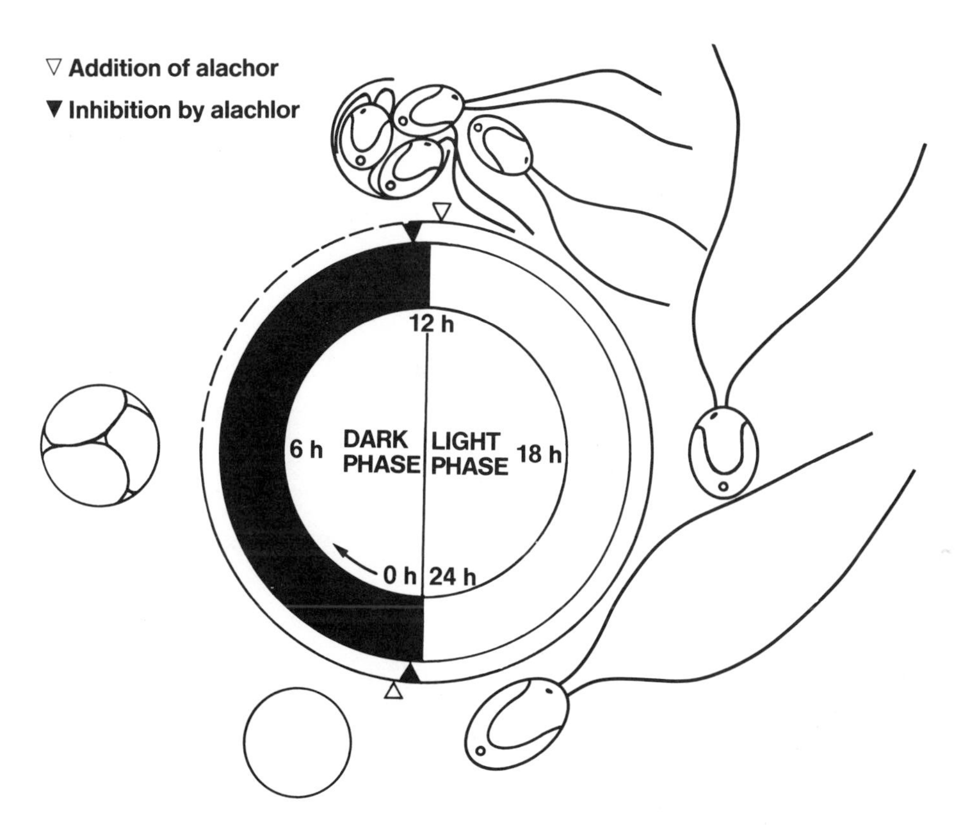

Figure 1. Chlamydomonas *growth and cell division during a 24-h cell cycle. In the cells, a cup-shaped chloroplast with a pyrenoid and the eye spot are seen. The two flagella provide locomotion.*

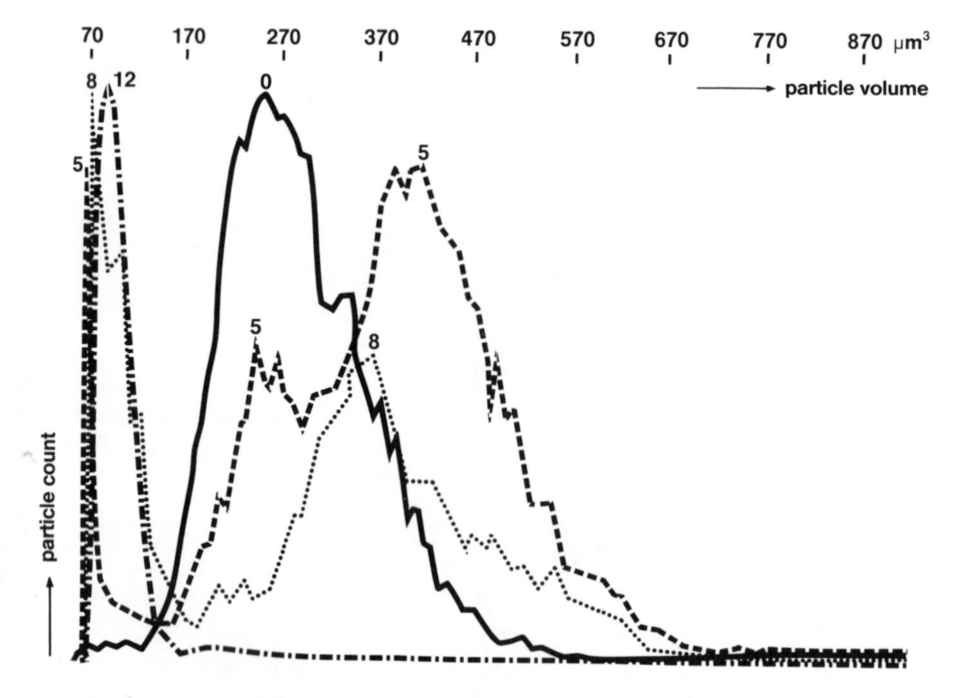

Figure 2. Size distribution of control Chlamydomonas *cell cultures during the dark phase. Numbers 0, 5, 8, and 12 mean after 0, 5, 8, or 12 hours in the dark.*

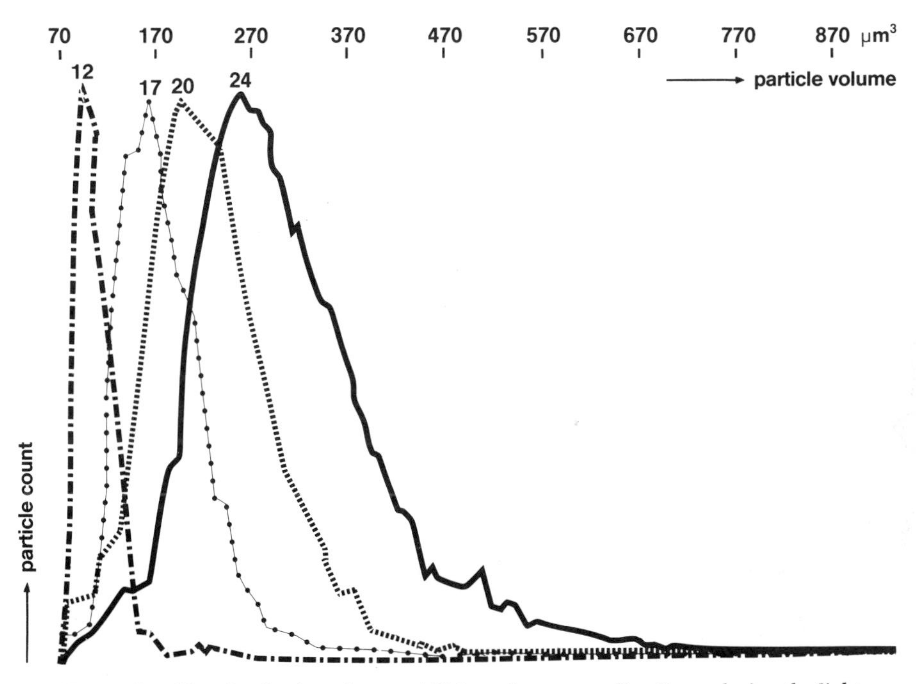

Figure 3. Size distribution of control Chlamydomonas *cell cultures during the light phase. Numbers 12, 17, 20, and 24 mean after 0, 5, 8, or 12 hours in the light.*

Table I. Common names, abbreviations (in parentheses), and chemical names of herbicides used in the present study

Common name	Chemical name
Alachlor	2-Chloro-2',6'-diethyl-*N*-(methoxymethyl) acetanilide
Amiprophos-methyl (=APM)	O-Methyl-O-(4-methyl-6-nitrophenyl-*N*-isopropyl-phosphorothioamidate)
Chlorpropham (=CIPC)	Isopropyl *m*-chlorocarbanilate
Fluridone	1-Methyl-3-phenyl-5-[3-(trifluoromethyl)phenyl]-4 (1*H*)-pyridinone
Metribuzin	4-Amino-6-*tert*-butyl-3-(methylthio)-*as*-triazin-5(4*H*)one
Nitrofen	2,4-Dichlorophenyl p-nitrophenyl ether
Paraquat	1,1'-Dimethyl-4,4'-bipyridinium ion
PCP	Pentachlorophenol
Trifluralin	α,α,α-Trifluoro-2,6-dinitro-*N*,*N*-dipropyl-*p*-toluidine

Table II. Action of the selected herbicides on *Chlamydomonas* growth and flagella regeneration and the most important underlying sensitive metabolic pathways or processes.

Herbicide	Metabolic pathway or process affected	*Chlamydomonas reinhardii* I_{50}-values (µM)	
		Growth	Flagella regeneration
Metribuzin	Photosynthetic electron transport	0.5	> 500
Paraquat	Photoreductions	0.1	34.1
Nitrofen	Photosynthesis	0.2	3.2
Fluridone	Carotenoid biosynthesis	6.1	121
PCP	Respiratory energy conservation	5.6	4.8
APM	Microtubule-controlled processes (cell growth, cell division, etc.)	0.3	0.3
Trifluralin		2.9	0.4
CIPC		1.0	52.4
Alachlor	Cell division	2.6	> 500

I_{50} = the concentration that inhibits by 50 %.

The sensitivity of the algal cells varied considerably among different tests. Thorough standardization was therefore necessary. The conditions applied for the estimation of the I_{50}-values in Table II (growth over 4 to 5 generations, starting with low cell densities) differed from those in the tests recorded with the Coulter Counter (growth over 2 generations, starting with higher cell densities). Moreover, duration of preculture of the cells in aqueous medium or on agar and prior duration of synchroneous growth were other parameters which were observed to change the sensitivity. However, controlling these factors did not entirely eliminate the variations in sensitivity. The concentrations applied in the recorded experiments were selected for optimum effectivity. In some cases, where the characteristics of the herbicidal action changed with the concentration, a lower and a higher one are recorded.

Effects of Herbicides Interfering with Photosynthesis

Metribuzin represents the herbicides that inhibit photosynthetic electron transport shortly after photosystem II (8). As expected, metribuzin does not inhibit cell division but inhibits cell growth in the light (Table III and Figure 4). Paraquat, which is toxic in the light through the photosynthetic reduction of oxygen to the superoxide anion (1), also does not affect cell division, but leads to rapid cell death in the light. The activity of nitrofen at low concentrations resembles that of paraquat, indicating activation or activity of the herbicide in the light (10). At higher concentrations, however, cell division is also affected. Fluridone, an inhibitor of carotenoid biosynthesis (11), neither interferes with cell division nor inhibits cell growth. However, the cells are bleached after 24 h and do not divide or grow further. Metribuzin and fluridone at 10 times higher concentrations did not show increased activity when compared with the lower concentrations.

Effects of Herbicides Interfering with Cell Division

Cell division is a complicated process, and many different primary interferences might eventually lead to the inhibition of cell division. One such primary interference that indirectly blocks cell division because the cells are rapidly killed is the uncoupling of oxidative phosphorylation, as demonstrated by PCP (Table IV). Many different inhibitory agents have been tested previously for their effect on *Chlamydomonas*, and a correlation between inhibitory mechanism and accumulating algal stage has been obtained (12, 13). In this study, four growth inhibiting herbicides with a more direct inhibitory action on cell division were selected.

Table III. Influence of herbicides that interfere with photosynthetic processes on growth and multiplication of *Chlamydomonas reinhardii*. The algae were kept in a synchronized 12/12 h light/dark cycle, with 0 h at the beginning of a dark phase.

Herbicide	Conc. µM	Average particle volume in μm^3 (h after start)				Increase of particle number (%)	
		0	12	24	36	0 - 14 h	24 - 36 h
Control	-	300	75	250	75	100	100
Metribuzin	0.5	↓	70	140	90	98	32
Paraquat	0.1	↓	95	220	90	100	25
	2.0	↓	90	70 †	-	107	0
Nitrofen	0.2	↓	90	160 †	-	68	0
	4.0	↓	210 [a]	200 † [a]	-	13	0
Fluridone	6.1	↓	70	250 [b]	170 [b]	100	13

† cells dead

a heterogeneous size distribution

b cells bleached

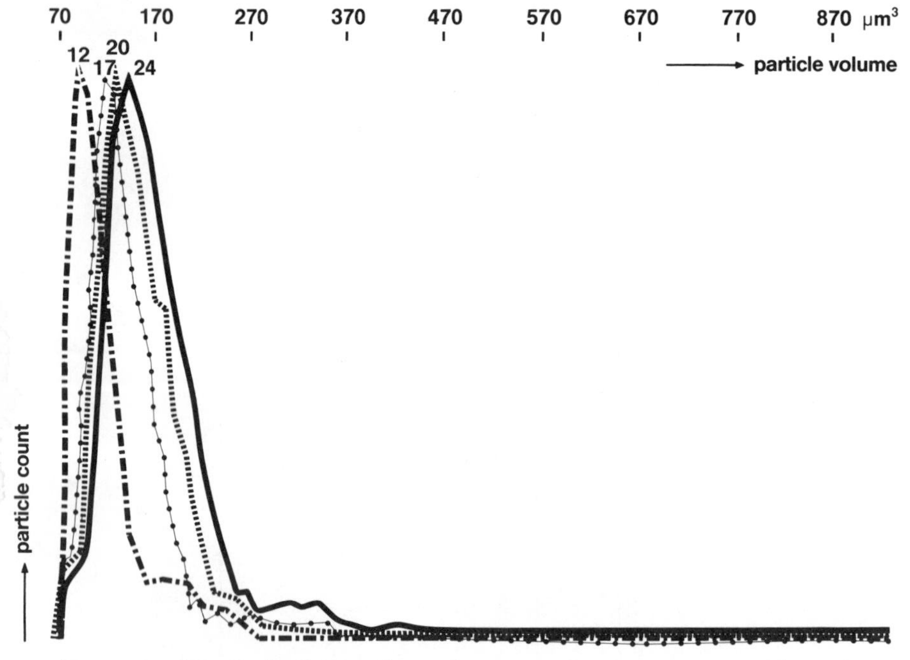

Figure 4. Size distribution of Chlamydomonas *cell cultures treated with metribuzin at time 0 (compare with Figure 3).*

Table IV. Influence of herbicides that interfere with cell division on growth and multiplication of *Chlamydomonas reinhardii*. The algae were kept in a synchronized 12/12 h light/dark cycle, with 0 h at the beginning of a dark phase.

Herbicide	Conc. µM	Average particle volume in μm^3 (h after start)				Increase of particle number (%)	
		0	12	24	36	0 - 14 h	24 - 36 h
Control	-	300	75	250	75	100	100
PCP	37	↓	300	270 + a	-	27	0
APM	6.7	↓	330 a	750 + a	-	0	0
Trifluralin	2.9	↓	90	400 a	90	27	100
	29	↓	300 a	600 a	500 + a	0	0
CIPC	19	↓	250 a	160 + a	-	30	0
Alachlor	2.6	↓	150 a	700 a	500 + a	42	0
	37		300 a	800 a	450 + a	10	0

+ cells dead

a heterogeneous size distribution

The herbicide APM blocks cell division very rapidly (Figure 5). Although the flagella were lost, a normal swelling reaction did not even occur at 8 hours into the dark phase. At the concentration applied, a few young (small) cells hatched at the end of the dark phase. After an additional 12 h growth in the light (Figure 6, trace 24), both small and large cells had grown considerably. Although some cells were dead at this time, others survived without division and enlarged during the second light phase, gaining a tremendous size. Cell volumes up to 8000 μm^3 have been measured in the microscope after 48 hours in the presence of 1.3 μM APM.

Trifluralin and CIPC are inhibitors of the microtubule system like APM (9). It may, however, be anticipated that they interfere with different molecular sites of action. Their effect on *Chlamydomonas* resembles that of APM in that they block cell division but not the growth of the cells (Table IV). The limited action of low trifluralin concentrations may be explained by the fact that trifluralin tends to adsorb to glass and other surfaces because of its low water solubility and, therefore, is removed from solution.

Alachlor does not interfere with the microtubular system. However, this herbicide also acts by inhibition of cell division (14). In *Chlamydomonas*, alachlor provoked rather peculiar effects: although normal cell division was initiated in the dark, few daughter cells were liberated (Figure 7). Instead, the cellular aggregates enlarged considerably during the following light phase, forming high volume particles (Figure 8). In the microscope, cellular aggregates of up to 2000 μm^3 have been measured.

Further Analysis of the Inhibition by Alachlor. Little is known about the herbicidal mode of action of alachlor and *Chlamydomonas* might, therefore, represent an interesting system for further studies. Figure 9 demonstrates that alachlor acts on cell division, but not on growth as measured by the increase in chlorophyll. Chlorophyll content and particle number, representing cell growth and cell division, respectively, show an increase only in the respective light or dark phase. When alachlor is added at the beginning of the dark phase, cell growth as measured by the chlorophyll content continues to increase normally for 24 h, whereas cell division (more precisely cell separation, Figures 7, 8) is blocked.

More information about the inhibition of cell division by alachlor was obtained from the experiment recorded in Table V. Stationary algae (grown to a high cell density where lack of light and nutrients limits further growth and division) were used when starting this experiment. Similar results were obtained when algae were taken from the logarithmic growth phase. The lower cell volume and cell multiplication factor are caused by these specific conditions. When alachlor is given at the

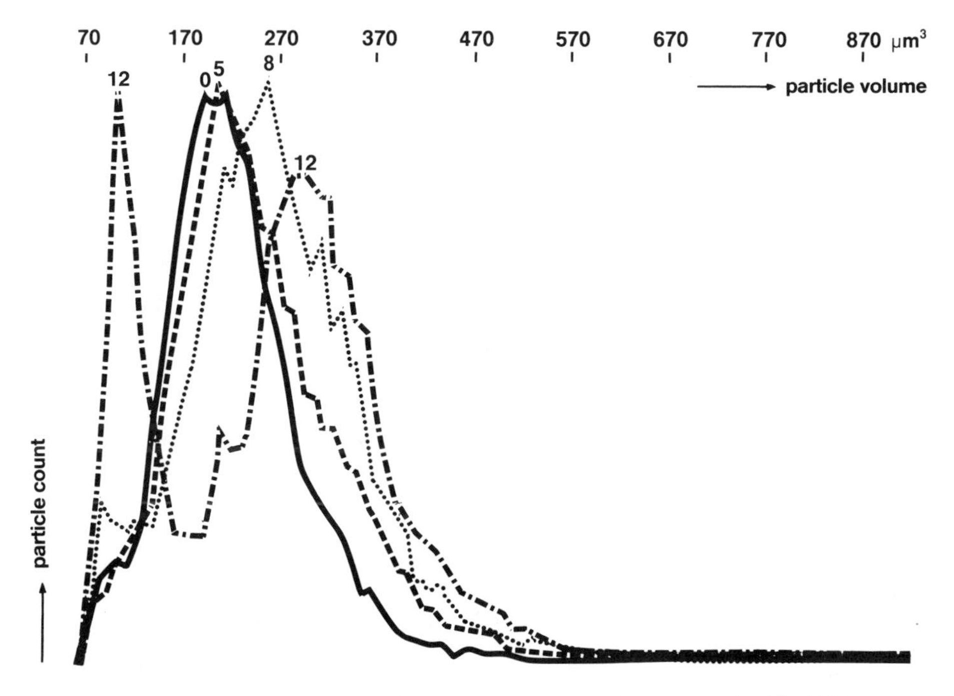

Figure 5. Size distribution of Chlamydomonas *cell cultures treated with APM at time 0 (compare with Figure 2).*

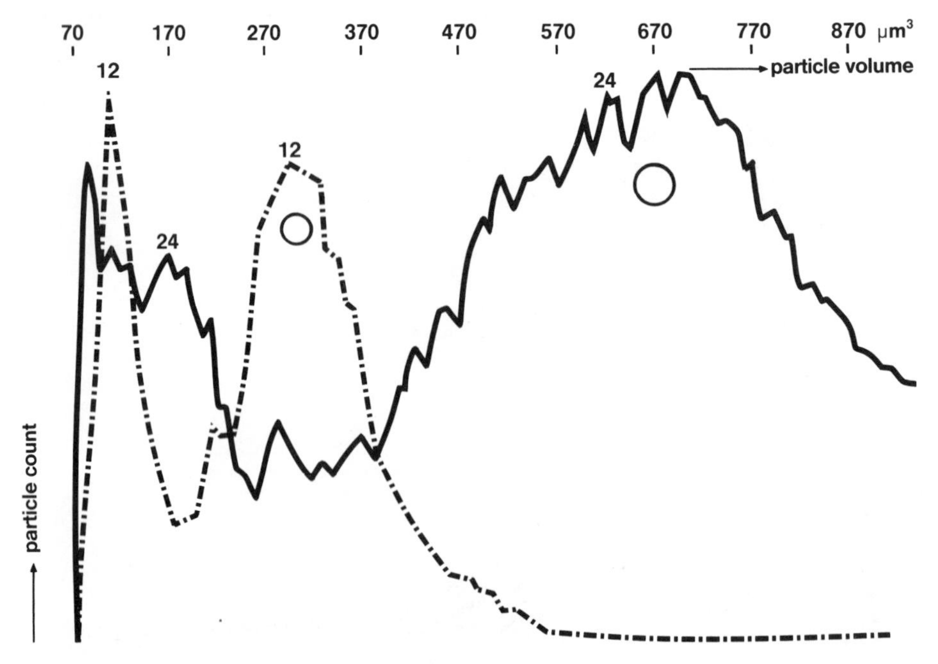

Figure 6. Size distribution of Chlamydomonas *cell cultures treated with APM at time 0 (compare with Figure 3).*

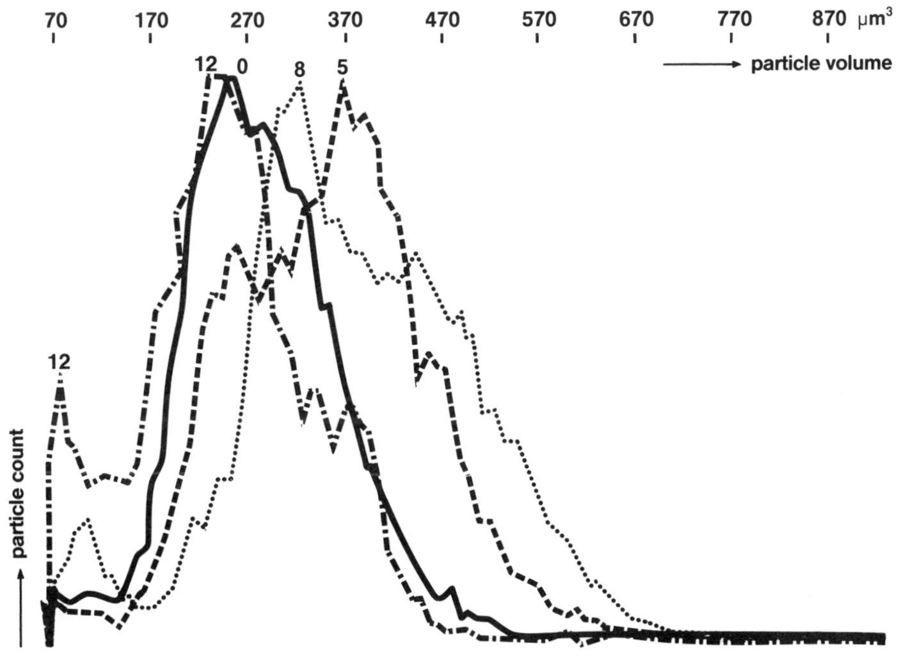

Figure 7. Size distribution of Chlamydomonas *cell cultures treated with alachlor at time 0 (compare with Figure 2).*

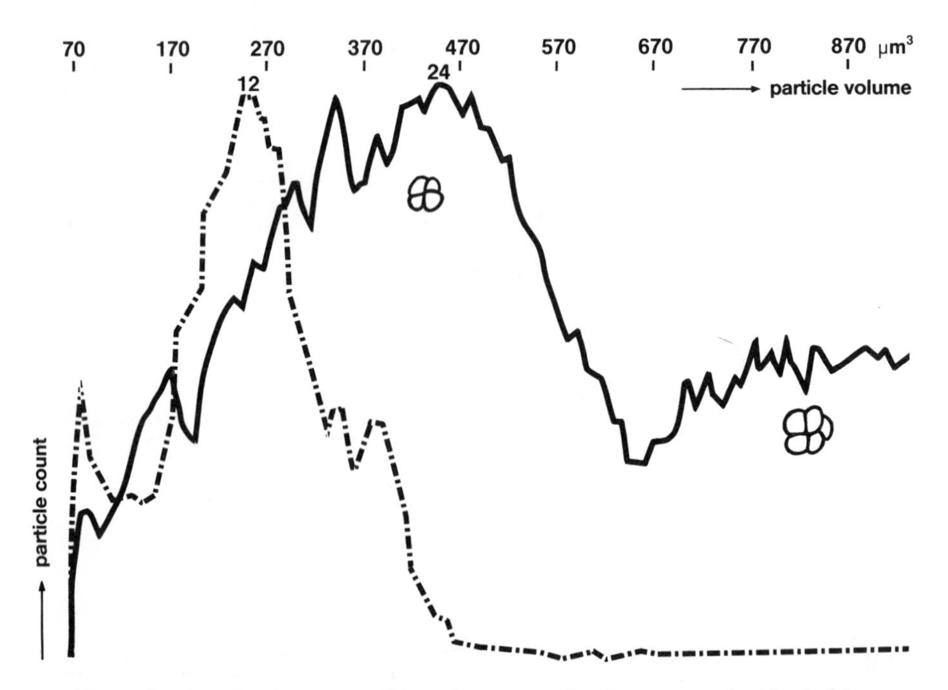

Figure 8. Size distribution of Chlamydomonas *cell cultures treated with alachlor at time 0 (compare with Figure 3).*

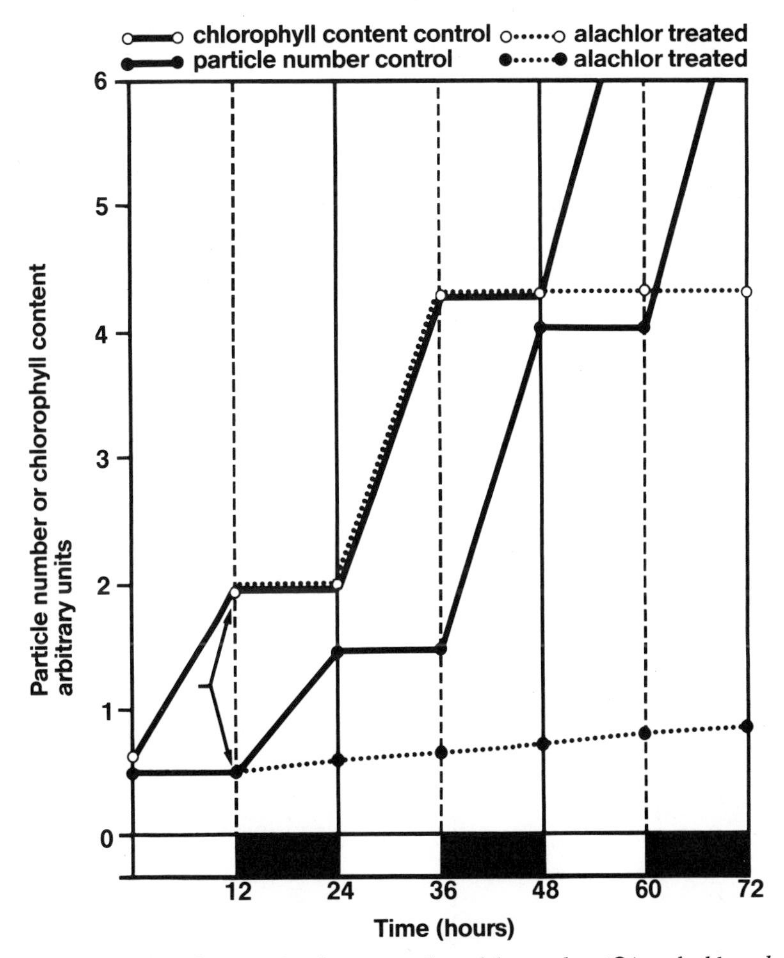

Figure 9. Idealized curves for increases of particle number (●) and chlorophyll content (○) per culture vessel, and for alachlor action (· · ·) on these increases. Alachlor application at arrow.

Table V. Action of 37 μM alachlor on *Chlamydomonas reinhardii* growth and multiplication when present for different time-intervals during the growth cycle. Inoculation was with stationary algae with a cell volume of 90 μm^3. Light phases were from 0 - 10, 24 - 34 and 48 - 58 h. C = Control, A = alachlor-treated.

Time-interval of alachlor-presence		Average particle volume in μm^3 after h						Increase of particle-number during time-interval in %		
h		8	24	32	48	56	72	0 - 30	30 - 54	54 - 72
Ø	C	181	84	156	92	183	115	268	184	210
0 - 72	A	200	440 [a]	230	220	-	-	10	30	0
9 - 72	A	150	220 [b]	580 [b]	400 [b]	-	-	60	0	0
0 - 9	A	160	230 [a]	500 [a]	100	200	100	30	690	220
24 - 72	A	210	80	150	310 [a]	170	150	310	20	40
0 - 24	A	190	440 [a]	270	100	200	110	0	70	250

[a] single cells without flagella

[b] aggregates of 2 - 5 cells

beginning of the first light phase (Table V, 0 - 72), no divisions occur during the following dark phase. However, when alachlor is added 9 hours later at the beginning of the dark phase (Table V, 9 - 72) the already familiar response is obtained: division occurs, but the daughter cells do not separate (compare Figures 7, 8). Alachlor, therefore, clearly is an inhibitor of cell division, albeit indirectly. Cell growth is not affected, i.e., the treated single cells (Table V, 0 - 72) or cell aggregates (Table V, 9 - 72) continue to enlarge until they eventually die.

When alachlor is removed from the algal suspension after 9 h (Table V, 0 - 9) by simple sedimentation and resuspension in fresh medium, no divisions occur in the first cycle. Alachlor, therefore, seems to block some metabolic process that is preparatory for cell division. The cells grow further and, after being released from the inhibition of cell division, divide extensively until the normal size of young cells is again attained. At that time, control and treated cultures are again equal, and alachlor has done nothing else but halted cell division during one cycle.

When alachlor is added at the beginning of the second light phase (Table V, 24 - 72), the inhibitory effect is the same as if given from the beginning. However, when alachlor is removed 24 h after addition, i.e., at the beginning of the second light phase (Table V, 0 - 24), the inhibitory effect is much stronger than for Table V, 0 - 9, and one entire growth cycle is deleted. Even the second cycle is partially inhibited under these conditions. The experiments (0 - 9) and (0 - 24) in Table V clearly demonstrate that the inhibition by alachlor is reversible.

Conclusions

The results demonstrate the usefulness of synchronized *Chlamydomonas reinhardii* cell cultures for herbicide mode of action studies. Inhibitory actions in the light (metribuzin) and toxic actions in the light (paraquat, nitrofen) are clearly discerned. Similarly, bleaching herbicides (fluridone) and respiratory uncouplers (PCP) may be classified from the algal response. Cell division inhibiting herbicides may be recognized as either direct (APM, trifluralin, CIPC) or indirect (alachlor) inhibitors. The algal responses do not allow for the formulation of a biochemical mode of action. However, the physiological type of response and the knowledge of the sensitive growth phase give valuable information.

In the case of alachlor, interesting new information has been obtained. The inhibition by alachlor is reversible and concerns a metabolic reaction or process that is required for normal cell division. The herbicide alachlor might specifically interfere with the regulation of cell division.

Acknowledgment. The experiments communicated here have been carried out by Mr. Jantzen.

Literature Cited.

1. Moreland, D.E. Ann. Rev. Pl. Physiol. 1980, 31, 597.
2. Kratky, B.A.; Warren, G.F. Weed Sci. 1971, 19, 658.
3. Sandmann, G.; Kunert, K.-J.; Böger, P. Z. Naturforsch. 1979, 34 c, 1044.
4. Hess, F.D. Weed Sci. 1980, 28, 515.
5. Amrhein, N.; Filner, P. Proc. Nat. Acad. Sci U.S.A. 1973, 70, 1099.
6. Mihara, S.; Hase, E. Plant Cell Physiol. 1971, 12, 225.
7. Quader, H.; Filner, P. Eur. J. Cell Biol. 1980, 21, 301.
8. Trebst, A. Proc. V. Int. Photosynthesis Congr. 1980.
9. Draber, W.; Fedtke, C. Proc. 4th Int. Congr. of Pesticide Chemistry (IUPAC) 1979; p. 475.
10. Fadayomi, O.; Warren, G.F. Weed Sci. 1976, 24, 598.
11. Bartels, P.G.; Watson, C.W. Weed Sci. 1978, 26, 198.
12. Howell, S.H.; Blaschke, W.J.; Drew, C.M. J. Cell. Biol. 1975, 67, 126.
13. Mihara, S.; Hase, E. Plant & Cell Physiol. 1978, 19, 83.
14. Deal, L.M.; Hess, F.D. Weed Sci. 1980, 28, 168.

RECEIVED September 29, 1981.

13

Use of *Chlorella* to Identify Biochemical Modes of Action of Herbicides

S. SUMIDA and R. YOSHIDA

Sumitomo Chemical Co., Ltd., Institute for Biological Science, 4-2-1 Takatsukasa, Takarazuka, Hyogo 665, Japan

Synchronized cultures of *Chlorella ellipsoidea* were used as a model system to identify biochemical modes of action of two herbicides, butamiphos (*O*-ethyl-*O*-(3-methyl-6-nitrophenyl)-*N*-*sec*-butyl-phosphorothioamidate, formerly coded S-2846), and chlorpropham (isopropyl *m*-chlorocarbanilate). Cell cycle studies showed that butamiphos inhibited cell division, whereas its effects on respiration and general biosynthesis were slight. Confirmatory experiments with onion root apices showed that mitosis was blocked at the metaphase and that the spindle apparatus was disrupted. Butamiphos seems to be a highly specific inhibitor of mitosis like colchicine. Chlorpropham inhibited cell division of a synchronized *Chlorella* culture. Respiration was not significantly affected. Protein synthesis was more sensitive to chlorpropham than the other biosynthetic processes tested, but a possibility remains that chlorpropham may directly inhibit mitosis without interfering with protein synthesis.

The higher plant is a multicellular system that consists of various organs such as leaves, stems, roots, and flowers. Comparatively, the unicellular alga is a simpler system. Taxonomically, the green alga is relatively closely related to the higher plant. Because of this relationship, unicellular green algae such as *Chlorella* have long been used by biochemists and plant physiologists as simple model systems for

0097-6156/82/0181-0251$05.00/0

higher plants, particularly to study the biochemistry of photosynthesis. Unicellular green algae are suitable to generate quantitative and clear-cut results under rigorously defined experimental conditions within a short period of time. Furthermore, the growth of unicellular green algae can be synchronized with relative ease, which provides a particularly useful system for a biochemical study of cell division in contrast to most higher plant tissue in which cell division is asynchronous. This paper describes the successful use of *Chlorella ellipsoidea* (abbreviated as *Chlorella*) as a simple model system to study biochemical modes of action of the herbicides, butamiphos and chlorpropham, which are inhibitors of cell division.

Butamiphos

Butamiphos is a phosphorothioamidate herbicide [*O*-ethyl-*O*-(3-methyl-6-nitrophenyl)-*N*-*sec*-butylphosphorothioamidate] formerly referred to as S-2846 (see Figure 1). Butamiphos controls annual weeds by preemergence application (1). Butamiphos acts on the root apex and causes it to swell noticeably. When *Chlorella* is treated with butamiphos, the cell population increase is inhibited. The average size of the treated cells becomes larger relative to the untreated controls (2). By assuming that swelling of the root apices of higher plants and the enlarged average size of *Chlorella* cells are caused by a common mechanism of action, we decided to use *Chlorella* as a model system to localize a biochemical site of action of butamiphos.

Previous research has shown that the cell population increase of *Chlorella* was inhibited 50% at about 2 μM of butamiphos and nearly 100% at about 6 μM (2). Effects of butamiphos on respiration and cell membrane permeability were examined. Butamiphos at 150 μM showed no significant effects on respiration as measured by oxygen consumption (2). This same concentration did not affect significantly cell membrane permeability as measured by leakage of radiolabeled intracellular materials (2). This means that these biochemical processes were not affected by butamiphos at a concentration 25 times greater than that which inhibited cell population increase nearly completely. Effects of butamiphos on biosynthetic processes were examined. Biosyntheses were measured by incorporation of an appropriate radiolabeled precursor into the respective fraction. At 30 and 150 μM, butamiphos showed slight effects on biosynthesis of cell wall material and RNA after a 60-min treatment (2). However, butamiphos had almost no inhibitory effect at 6 μM, the concentration at which the cell population increase was almost completely inhibited. Effects of 24-h treatment with 6 μM of butamiphos on biosynthetic processes were studied. Butamiphos had no inhibitory effects on biosyntheses of protein, lipid, cell wall polysaccharide, RNA, and DNA on a per cell basis. Instead, per cell biosyntheses tested were stimulated 160 to 300% (2). During the course of the investigation, a *Chlorella* culture was accidentally contaminated by bacteria. Bacteria were actively propagating at the butamiphos concentration where *Chlorella* cell population increase was completely inhibited. In order to confirm this observation, effects of butamiphos on a pure culture of *Escherichia coli* was examined as a model bacterium. *E. coli* was more than 150 times more resistant to butamiphos than *Chlorella* (2). Under these circumstances, it seemed possible that butamiphos acted on a process peculiar to eukaryotic cell division.

In order to further test the effect of butamiphos on *Chlorella*, a synchronized culture was used. In the cell cycle of *Chlorella*, a young cell grows in biomass. After nuclear divisions, the cell divides to give an even number of daughter cells. The daughter cells repeat the same cycle. It takes about 24 h for one complete cell cycle. In a synchronized culture, individual cells in the whole population are synchronized with respect to their cell stages so that they grow in biomass and undergo cell division approximately simultaneously. Synchronization can be accomplished quite easily by culturing *Chlorella* at 24 C in an 18-h light and a 6-h dark period (3).

Using a synchronized culture, the effect of butamiphos on cell division was studied (Figure 2). In the untreated control, the cell number was constant until cell division started at 18 h and then the cell number increased at an abruptly rapid rate. The total volume of the cell population increased continuously throughout the cell cycle as the individual cells grew in biomass. The average volume per cell gradually increased until cell division started, and then it became abruptly smaller. This pattern is characteristic of a synchronized culture. If butamiphos specifically inhibits the cell division process without affecting other synthetic processes such as respiration and photosynthesis, then the cell number should be constant throughout the cell cycle, the total cell volume of the population should increase continuously (as in the case of the untreated control), and

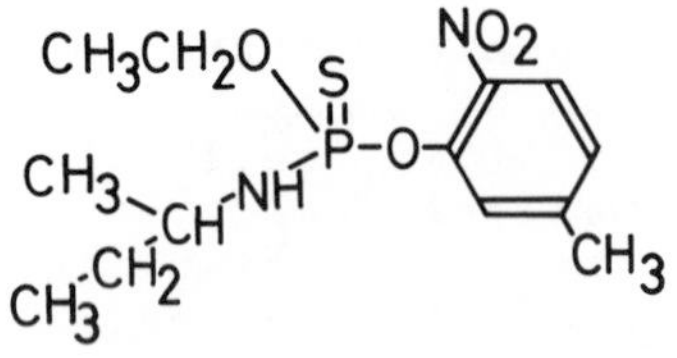

BUTAMIPHOS (S-2846)

Figure 1. Chemical structure of butamiphos.

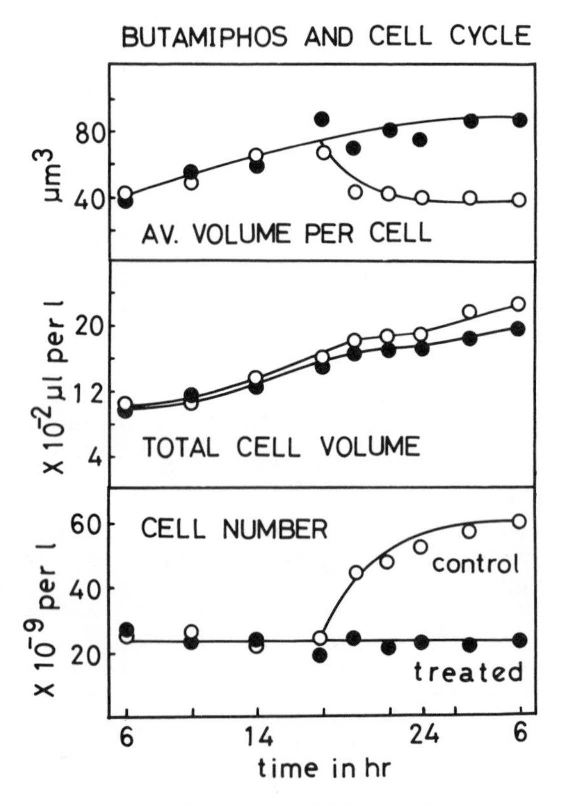

Figure 2. Effect of butamiphos on cell division of Chlorella *grown in synchronized culture. Butamiphos in methanol was added to the culture after 6 h. Aliquots were taken for the measurement of cell number and packed cell volume at various time intervals. The dark period was between the 18th and the 24th h. Key: ○, untreated control; and ●, butamiphos-treated (30 µM).*

the average volume per cell should increase in a continous manner. As shown here, the butamiphos-treated cells behaved exactly as expected (Figure 2). Effect of butamiphos on average DNA content per cell was examined by using synchronized cultures. The average DNA content per cell treated at 30 μM for 24 h as measured by a modification of Schmidt-Thannhauser's procedure (3) became 3.9 times greater than that of the untreated control. This seemed reasonable because one untreated cell yielded 4 daughter cells after division. This indicates that butamiphos did not act on the DNA duplication step. It seemed likely that butamiphos acted on a certain process occurring after DNA synthesis. If butamiphos acts on a specific step in the cell division process, then there would exist a butamiphos-sensitive period that should be transient in nature. Treatment of cells with butamiphos before this period would block the cell cycle, but the treatment after this period would allow the cycle to proceed until the cells come to the butamiphos-sensitive period in the next cycle. In Figure 3, the effect of butamiphos treatment at different times in the cell cycle is given. When cells were treated with butamiphos at the 6th, 10th, or 15th h, cell division was completely inhibited. When cells were treated at the 17th h, about 10% of the cells divided. When treated at the 18th h, more than 50% of the cells divided. This indicated a butamiphos-sensitive period ended around the 17th to 18th h. According to electron microscopic studies by Takabayashi *et al.* (4), mitosis of *Chlorella ellipsoidea* starts around the 16th h. This approximately coincides with the butamiphos-sensitive period.

The above observation led to an experiment to ascertain if butamiphos acted directly on mitosis. Unfortunately, *Chlorella* was not an appropriate system to study mitosis by optical microscopy. Therefore, it seemed desirable to use a higher plant system. Onion root tip tissue was selected for this purpose. A typical division figure found in butamiphos-treated tissues was that, unlike the untreated cells, chromosomes were not aligned along the equatorial plate, but dispersed in the cytoplasm, giving rise to a so-called arrested metaphase (5). It is well known that colchicine inhibits mitosis at the metaphase. Butamiphos-treated onion cells had the same pattern of mitotic inhibition as colchicine-treated cells used as a reference (5). Butamiphos-treated tissues were compared with the untreated control with respect to percentage distribution of mitotic cells in each mitotic

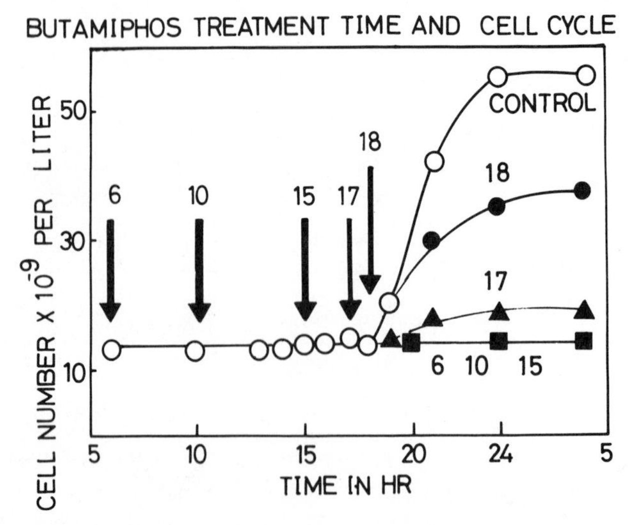

Figure 3. Effect of butamiphos (30 μM) on cell division of Chlorella *grown in synchronized culture and treated at different time intervals. Butamiphos was added at one of the times indicated by arrows. Aliquots were taken for the measurement of cell number at various time intervals. The dark period was between the 18th and 24th h. Key: ○, untreated control; ●, treated at the 18th h; ▲, treated at the 17th h; and ■, treated at the 6th, 10th, or 15th h.*

phase. In the untreated control, cells in the prophase plus metaphase accounted for about 60%, and those in the anaphase plus telophase for about 40%. On the other hand, in the butamiphos-treated tissues, cells in the prophase plus metaphase accounted for more than 99%, and anaphase plus telophase cells were nil (5). The reduction in percentage of anaphase plus telophase cells was attributed to the increase in percentage of arrested metaphase cells. It is known that if colchicine-treated metaphase cells are subjected to centrifugation at 1500 x g for 5 min, chromosomes move in the direction of centrifugal force, because spindle fibers are broken (6). Butamiphos-treated cells were subjected to this centrifugation experiment. As expected, chromosomes moved in the direction of centrifugal force suggesting that butamiphos inhibited mitosis at the metaphase by disrupting spindle fibers (5).

Wanka reported that colchicine prevented cell division in synchronized culture of a *Chlorella* species (7). The cells became polyploid with regard to their DNA content. RNA and protein syntheses were not inhibited. The target of colchicine action was considered to be short developmental stages during successive nuclear duplication steps. Compared with our results, there seems to be a striking similarity in time and mode of action between butamiphos and colchicine, although their chemical structures are entirely different. Similarity in mode of action was also observed in the onion root apex where butamiphos blocked mitosis at the metaphase by disrupting the spindle apparatus. The overall results indicate that butamiphos is a highly specific inhibitor of mitosis. It does not act on respiration, membrane permeability, or major biosynthetic processes.

Chlorpropham

Chlorpropham is isopropyl *m*-chlorocarbanilate as shown in Figure 4. A number of reports have appeared concerning biochemical responses of higher plants to chlorpropham. Chlorpropham was reported to inhibit mitosis in onion root (8). Mann *et al.* reported that chlorpropham inhibited protein synthesis in barley coleoptiles and sesbania hypocotyls (9). Moreland *et al.* observed that chlorpropham inhibited the biosynthesis of RNA and protein in soybean hypocotyls and RNA synthesis by corn mesocotyls (10). Rost and Bayer reported that synthesis of protein was reduced before DNA and RNA synthesis were inhibited in pea root (11). Lotlikar *et al.* found that chlorpropham inhibited oxidative phosphorylation in cabbage mitochondria (12). St. John reported that chlorpropham reduced the *in vivo* level of ATP in *Chlorella sorokiniana* (13). The investigators conducted these studies using different plant tissues and different concentrations of chlorpropham. Because chlorpropham is known to inhibit cell division, it is necessary to identify the relative contributions of these biochemical effects to inhibition of cell division. Does chlorpropham directly affect oxidative phosphorylation, which leads to inhibition of cell division? Does it directly affect major biosynthetic processes, which lead to inhibition of cell division? Or, does it directly act on a cell division process without affecting energy metabolism and biosyntheses? In order to make direct comparisons among these biochemical effects of chlorpropham, *Chlorella* was used again as a model system.

Effects of chlorpropham on the cell cycle was studied (3) (See Figure 5). Cell population increase was completely suppressed by chlorpropham at 14.4 µM which was 3.6 times the LC_{50} concentration. Average cell volume became comparatively larger after one complete cell cycle. Average DNA content per cell was increased 3.0-fold (3). These observations indicate that chlorpropham directly inhibited cell division in *Chlorella*.

Effects of chlorpropham on respiration were studied (3). Oxygen uptake was not significantly affected by chlorpropham at a concentration up to 47 µM which was about 40 times the LC_{50}. Effects of chlorpropham on biosyntheses of protein, cell wall polysaccharide, RNA, and lipid were examined (3). At 47 µM, biosyntheses of protein and cell wall polysaccharides were substantially inhibited. RNA synthesis was not affected. At 4.7 µM, a slight inhibition of protein synthesis was observed, but an inhibitory effect on cell wall synthesis was not observed. Stimulatory effects on lipid synthesis were erratic, and, therefore, were not pursued further. Effect of chlorpropham on protein synthesis was examined at lower concentrations (3). At 2.4 µM or lower, biosynthesis of protein was not inhibited even after a 24-h treatment. Cell population increase was inhibited 50% at 1.3 µM. Among the biosynthetic processes studied, protein synthesis was the most sensitive to chlorpropham. However, considering the LC_{50} mentioned above, it would be premature to conclude that protein synthesis is the primary site of action of chlorpropham.

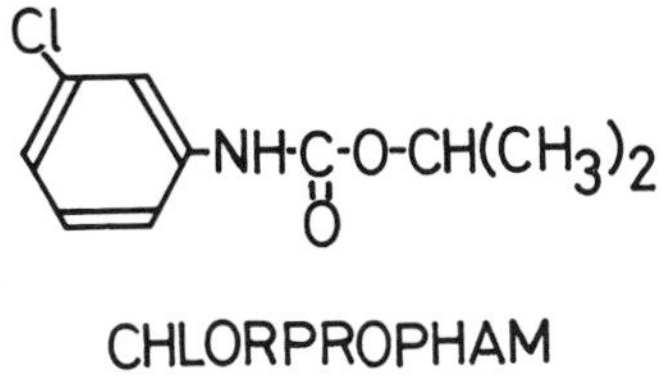

Cl

$NH-C(=O)-O-CH(CH_3)_2$

CHLORPROPHAM

Figure 4. Chemical structure of chlorpropham.

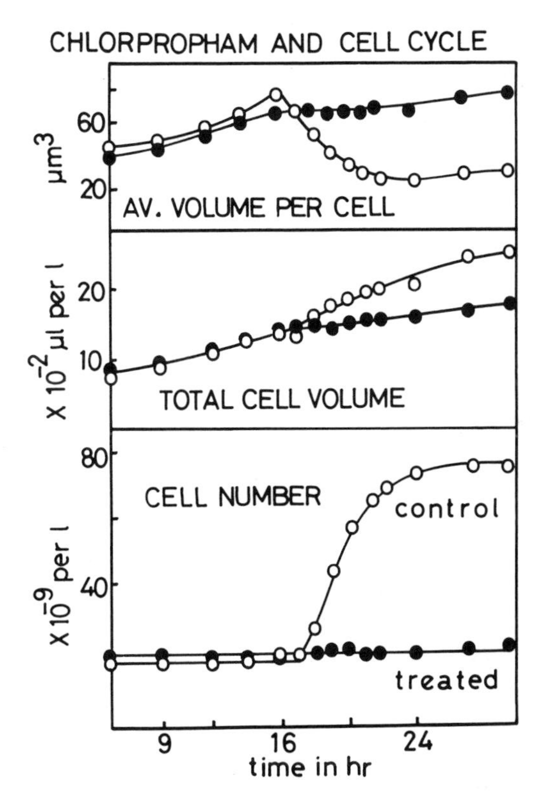

Figure 5. Effects of chlorpropham on cell division of Chlorella *grown in synchronized culture. Chlorpropham in methanol was added after 5 h. Aliquots were taken for the measurement of cell number and packed cell volume at various time intervals. The dark period was between the 12th and 18th h. Key: ○, untreated control; and ●, chlorpropham-treated (14.1 μM).*

Conclusions

Synchronized cultures of *Chlorella ellipsoidea* were used as a model system to study biochemical modes of action of butamiphos and chlorpropham, both of which are inhibitors of cell division. Cell cycle studies showed that butamiphos was a highly specific inhibitor of mitosis like colchicine. As for chlorpropham, protein synthesis was inhibited most sensitively, but a possibility was suggested that chlorpropham could directly inhibit mitosis without interfering with protein synthesis. It will be interesting to ascertain how plant mitosis is affected by herbicidal organophosphoramidates and carbamates at the molecular level in more detail as will be studies involving insecticidal organophosphates and carbamates.

Acknowledgements

The authors wish to express their thanks to Sumitomo Chemical Co., Ltd. for permission to publish this work.

Literature Cited

1. Ueda, M. *Japan Pestic Info.* 1975, *23*, 23-5.
2. Kohn, G. K. Ed.; "Mechanism of Pesticide Action"; American Chemical Society: Washington, DC., 1974; p. 156-68.
3. Sumida, S; Yoshida, R.; Ueda, M. *Plant Cell Physiol.* 1977, *18*, 9-16.
4. Takabayashi, A; Nishimura, T.; Iwamura, T. *J. Gen Appl. Microbiol.* 1976, *22*, 183-96.
5. Sumida, S.; Ueda, M. *Plant Cell Physiol.* 1976, *17*, 1351-4.
6. Lingnowski, E. M.; Scott, E. G. *Weed Sci.* 1970, *20*, 267-70.
7. Wanka, F. *Arch. Mikrobiol.* 1965, *52*, 305-18.
8. Ashton, F. M.; Crafts, A. S. "Mode of Action of Herbicides"; John Wiley & Sons: New York, 1973; p. 200-20.
9. Mann, J. D.; Jordan, L. S.; Day, B. E. *Weeds* 1965, *13*, 63-6.
10. Moreland, D. E.; Malhotra, S. S.; Gruenhagen, R. D.; Shokraii, E. H. *Weed Sci.* 1969, *17*, 556-63.
11. Rost, T. L.; Bayer, D.; *Weed Sci.* 1976, *24*, 81-7.
12. Lotlikar, P. D.; Remmert, L. F.; Freed, V. H. *Weed Sci.* 1968, *16*, 161-5.
13. St. John, J. B. *Weed Sci.* 1971, *19*, 274-6.

RECEIVED September 14, 1981.

INDEX

INDEX

D

E

H

T

U

V

W

Z

Jacket design by Kathleen Schaner.
Production by Katharine Mintel and V. J. DeVeaux.

Elements typeset by Service Composition Co., Baltimore, MD.
Printed and bound by The Maple Press Co., York, PA.